TWIDDERS

By

Anita Holmes

OZARK
MOUNTAIN
PUBLISHING

PO Box 754, Huntsville, AR 72740

800-935-0045 or 479-738-2348

fax 479-738-2448

For permission, serialization, condensation, adaptions, or for our catalog of other publications, write to Ozark Mountain Publishing, Inc., P.O. box 754, Huntsville, AR 72740, ATTN: Permissions Department.

Library of Congress Cataloging-in-Publication Data

Holmes, Anita, 1950-

    TWIDDERS, by Anita Holmes

True accounts of time slips and distortions..

1. Time-slips   2. Time Distortion   3. Quantum Physics   4. Metaphysics

I. Holmes, Anita, 1950-   II. Time-slips   III. Metaphysics

IV. Title

Library of Congress Catalog Card Number: 2010939624

ISBN: 978-1-886940-116

Cover Art and Layout: www.enki3d.com

Book set in: Times New Roman, BernhardModE

Book Design: Julia Degan

Published by:

**OZARK
MOUNTAIN
PUBLISHING**

PO Box 754
Huntsville, AR 72740
www.ozarkmt.com
Printed in the United States of America

DEDICATION:

To my grandchildren—MaKinzie, Analisia, Makaylee, Logan, Brynlee, Asher, Kailen, Kellen, Brayden, Kamden, Brianna, Maranda, and Sydney. Through your eyes, our world is always wondrous and surprising.

ACKNOWLEDGMENTS:

I would like to thank the many, many individuals, like Terry Burgoyne, who so courageously shared their time-slip experiences, in spite of possible ridicule or incredulity. Special thanks to all of the authors and website directors for so graciously granting permission to include TWIDDERS accounts from their publications; Sally Heuer with Llewellyn Worldwide, Tim Swartz, Herbie (J.H.) Brennan, Tom Slemen, Stephen Wagner of paranormal.about.com, Deena Budd of BellaOnline, obiwan.@netcom.com, AboveTopSecret.com and mercuryrapids. Last but not least, my unending gratitude to the remarkable crew at Ozark Mountain Publishing.

Table of Contents

Chapter 1.

Huh?

For teachers, October's a blast! Halloween gives us license to go wild and crazy on everything from bulletin board design to language arts units.

I go out of my way to present my high school students with a knock-your-socks off English theme for those four short but wacky weeks in the tenth month of the year.

A few years ago I decided we'd focus on allegedly true ghost stories from our home state, Alaska. I checked out a truckload of books from the Fairbanks Public Library and spent the weekend knee deep in lurid tales of long dead miners standing guard over slag heaps, mushers still traversing winter trails with faithful dogs towing phantom sleds, and even sightings of sunken ships on the Yukon.

One account in particular piqued my interest. Rather than your quintessential ghost story (typically featuring folks catching a filmy slice of the past, i.e. "residual" hauntings), it was an interactive experience with one VERY live figure from decades earlier.

While the volume hosting this account is no longer part of the local library collection, I'll never forget the tourist's tale.

In the 1990's, a Washington state couple (let's call them the Smiths) took the cruise of a lifetime, sailing up Alaska's lower leg, exploring fjords and legendary far north outposts like Skagway.

Halfway through the cruise, the ship berthed overnight in Anchorage. The Smiths joined another couple for a hurried tourist trip through Old Town. At one point, they entered the historic Anchorage Hotel and spent time and money at the gift shop. (Technically, the structure housing the hotel is the 1936 annex to the original lodging. In 1989 the neglected building was brought back to life, and is now listed on the National Register of Historic Places.)

It was here that Will Rogers stayed, just two days prior to his fateful flight to Barrow. And here that Mrs. Smith would have a remarkable experience.

Still mulling over gifts for the grandchildren, she excused herself to use the restroom. Her husband and their friends were waiting in the foyer when she made her way back.

She caught up with them and rapidly launched into a description of the ladies' room . . . how the hotel had gone totally retro by re-creating it in an exact replica of the facility in 1936—right down to the white ceramic spigots, quaint lighting, linen hand towels, and old-fashioned toilet stalls with oak paneling. She said that while she was washing her hands, a gal dressed up as an old-fashioned hotel maid came in and handed her a clean towel.

Mrs. Smith was so enamored of this unexpected touch of bygone elegance to the now-modern hotel that she insisted her lady friend come take a look. Except when they entered the facility, it was as sleek and modern as the rest of the hotel, complete with recessed electric lights, metal stalls, and a blow dryer for hands.

Mrs. Smith managed to not pass out. To this day she staunchly asserts she saw, felt, used, and interacted with a room from the past!

Huh?

Reading this experience as told by a distinctly sincere and apparently sane woman sparked my interest in the phenomenon known by many names: time-slips, time-warp or temporal displacements, time jumps, time travel, time fabric ruptures, time ribbons-or strings-or yarn, time storms, time slippages, time tears, alternate realities, dimensional leap-frogging, etc. I lump them all under the acronym TWIDDERS (a jumble of "time/warp/displacement" with a pinch of "slip" thrown in). By its spontaneous nature, a time-slip, or TWIDDER, is differentiated from someday-through-human-engineered-means time travel.

As the years have gone by, my files have filled with TWIDDERS accounts. Not all of the accounts I've come across were believable. The ones I took the trouble to hang onto struck me as being recounted by credible individuals. They've popped up in the most unexpected places: biographies of famous folk, written reminiscences, histories, websites, compilations of near death experiences, and published journals.

But as the stack of accounts thickened, the nature of the experiences broadened. By far the largest category logs people's treks back in time. In addition, I found accounts of time inexplicably speeding up, lost locations, time slowing down, leap frogging time, lost time, found time, different locales, instant replay, and rarest of all, alternate realities.

The common thread in these narratives was a shift in time and/or location that is contrary to our understanding of reality.

Or . . . maybe not . . .

I had a cabinet full of TWIDDERS accounts; I wanted to know if there was a scintilla of evidence that could lend them credence.

Science seemed the place to begin research. In particular, physics. Physics is the study of matter and energy. Since TWIDDERS deals with people, places, and times, surely the realm of physics might provide an explanation.

Wrapping our brains around much of the theory that may one day explain how TWIDDERS can occur is a mind-boggling proposition. It all begins with REALLY BIG THINGS and REALLY SMALL THINGS. And the physics that deal with each.

The theories of relativity and quantum mechanics revolutionized physics in the twentieth century. Pretty much all known physics is now built on these two constructs, which deal with big and small things, respectively.

Huh?

For just a moment, let's step back and begin with a really dumb thing: Gregg Eats Watermelon Sandwiches. (Yeah, well, there's no accounting for taste.)

Actually, I use that bizarre sentence to help students remember the four known forces of nature: Gregg = Gravity, Eats = Electromagnetism, Watermelon = Weak nuclear force, and Sandwiches = Strong nuclear force. Of the four, gravity is the weakest.

Gravity exists between all objects and can act over long distances. For example, it holds Earth in orbit around the sun and all the stars together in our galaxy—the more massive the object, the more impressive the gravity. And yet it's the weakest of the four forces. Go figure.

Albert Einstein introduced the theory of relativity about a century ago. Among other things, this theory addressed the concept of gravity. There are actually two parts to this remarkable model: 1) a special theory (published in 1905) that deals with the structure of space-time (more on this later); and 2) a general theory (published in 1916) that grapples with gravity. These theories have many surprising consequences, like time dilation, length contraction, warped space, and the concepts of not space and time as separate entities but as one and the same—space-time.

Quantum Mechanics (QM) deals with the other three known forces. Physicists are now trying to devise the physics version of the Holy Grail, a unified theory bridging the disparate findings of relativity and particle physics. Some think superstring or M-theory may be the ticket.

The theory of relativity is a gem at explaining how big bits of matter work and the concept of a fourth dimension. Yet it's all but useless in describing how really, really, really small things (matter on a sub-atomic level) work. Enter quantum mechanics, stage left, with a flourish.

Unless particle physics (the field addressing QM) is your

cup of tea, you have to be really desperate to immerse yourself in books dedicated solely to complex discussions of quarks, gauge bosons, Rindler transformations, leptons, dark energy, branes, the bulk, supersymmetry, string theory, the Standard Model, gravitons, the Planck scale, probability amplitudes . . . are you ready for some crumpets with your tea yet??? Even as I write this paragraph, my right eye starts twitching, and I have to grab for chocolate, because I did reach that point and turned to just such books.

Huh?

As part of my search to make sense of just how TWIDDERS could occur, I spent the better part of a year slogging through every science tome the local bookstores and libraries proffered. There were times when eating sawdust was more appealing than opening up to the next chapter, but, by golly, I did find germane developments in physics that enhance our understanding of reality and might just provide insight into TWIDDERS accounts. Plus, they draw on the wisdom that both relativity and QM have provided to-date. In particular, developments focus on brane (membrane) worlds, multi-verse theory, and the Now. Consequences of these current explanations of reality are fascinating.

But first things first; let's take a look at some of the remarkable ramifications of good ole relativity.

RELATIVITY

Now stick with me here while I take a quick trip to my personal past the old fashioned way—by fallible memory.

The year is 1966, and I'm a high school sophomore in an overcrowded geometry class at Wilson High in Portland, Oregon. In an effort to bolster my B-, I read a teacher-recommended math book. (By submitting a book report, my sagging B was brought up. Barely.)

But that book made such an impression on me that I still remember it. It was entitled *Relativity for the Layman* (Coleman,

1954), and it was just that. The author took a mind-boggling topic, Einstein's theory of relativity, and through the use of every day examples, made understandable to a short-sighted fifteen year-old many of the repercussions of this ground-breaking theory. (By the way, old copies of this book are still available. The author was James A. Coleman, an Englishman.)

One analogy in particular has remained with me. Coleman was helping the lay reader to understand the concept of higher dimensions.

In mathematics, the dimension of a space is roughly defined as the minimum number of coordinates needed to specify every point within said space. When I introduce graphing to students, we learn about 2 dimensional graphs, with the horizontal x line and the vertical y line. Points on the graph require two number coordinates—an ordered pair—to describe them. Shapes in a two coordinate or two-dimensional world (length and height) appear flat. Duh.

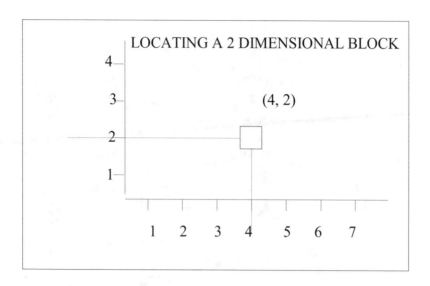

LOCATING A 2 DIMENSIONAL BLOCK

(4, 2)

Coleman (1954) had the reader imagine a piece of paper. The paper represented a two dimensional "world" (height, width, NO depth). Now imagine a stick figure—let's call him Tom—on the paper, heading east (as we view the paper, he's off at a nice clip to our right).

Tom
on
his
trek
East

Worlds without end, Tom cannot flip himself over and head west. To do so, he'd have to step into the third dimension (depth). Since he lives in a two-dimensional world, such a thing would be impossible! He can't comprehend such a thing! It'd fly in the face of *his* reality!

While we, superior beings that we are on our perch in the third dimension, can simply reach down, flick Tom's paper world over and voila! He'd be heading west. In Tom's eyes, it would be nothing less than a miracle!

Now imagine yourself in a closed room—no windows, no door. You walked in of your own volition through a single open side. (Why? Someone told you the world's biggest box of See's chocolate was in the far corner and you went for it! Or something like that. Just work with me here.)

At any rate, as soon as you walk in, the scumbag who lied to you about the candy seals you in. There's absolutely no way out! Well, that is, in a three dimensional world (height, width, AND depth).

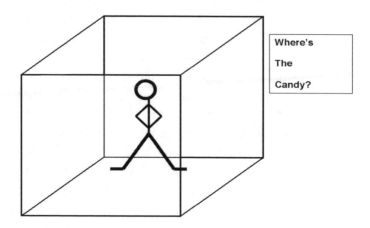

For an entity in the fourth dimension, it would be child's play for them to pick you up and set you on the floor outside your closed room. Intact. Impossible! Incomprehensible! Such a thing flies in the face of *our* reality!

And it does for three-dimensional beings. But not for those residing in the fourth dimension. No way, you say. *Way!*

In a nutshell, just because we can't comprehend something, doesn't mean it can't happen. By "happen," I mean through operating within the parameters of the laws of the universe. We're not talking magic here but simply the incredible ramifications for reality as described by modern physics theory. These outcomes have been in place and functioning in our universe for ages untold, but we are just now beginning to understand them. And some of these we rarely, if ever, will consciously experience—such as TWIDDERS.

Here's another example of a rarely experienced outcome. General relativity equations give birth to the existence of wormholes—tubular shortcuts in space-time—that may transport you really fast from one spot in space to another. Some speculate that wormholes could also be capable of whisking you to different times. Both ends of a wormhole could be intra-universe (existing in the same universe) or inter-universe (existing in different universes and serving as a connecting passage between the two.) *Note: Could wormholes be the answer to disappearing socks in the dryer? In some parallel universe are there folks trying to make sense of the multiplying-sock-in-the-dryer phenomenon???*

Consider yet another mind-bending revelation from relativity. At this point in time, our understanding is that only massless matter, like photons, can attain the speed of light. Should anything like, say me, manage the purportedly impossible and achieve that speed, I would CEASE TO AGE. (Would that this had happened fifty pounds and twenty years ago!)

Which brings us to special relativity's big contribution to reality; spacetime. Philosophers and scientists have oft wrestled with the issue of what's the true nature of space and time. Our minds have evolved a process of placing objects into a model of space and time, but to what extent is this model an accurate one of the real world?

Newton took it for granted that space and time existed as part of the humble stage on which the laws of physics play out their roles. Einstein's theory of special relativity unified space and time into the single super star geometry of SPACETIME.

In *Essence of Time*, Z. Goonay (2006) shared Minkowski's theory. Hermann Minkowski, a German mathematician, seconded Einstein's findings mathematically. He calculated a four-dimensional space, since known as "Minkowski space-time," in which time and space are intermingled in a four dimensional spacetime. He further showed mathematically that the whole of our past and the future must meet, can meet, always and forever do meet at a single point.

Everything, all of space and all of time exists all at once everywhere.

Huh?

I know. We all live by the clock. Clocks track the passage of time. While in the fourth dimension the time/aging process may halt, here in the third dimension time wreaks havoc on our bodies.

In T. Folger's article, *Newsflash: Time May Not Exist* (2007) physicists dubbed this the "problem of time." It may be the biggest temporal conundrum we're faced with, but not the only one. Vying for second place is this strange fact: on a quantum level, there is no past or future while on a macro level, the laws of physics can't explain why time always points to the future. All of the laws—whether Newton's, Einstein's, or QM's—would work equally well if time ran backward. Yet as far as we can tell, time is a one-way process; it never reverses even though no laws restrict it.

Whether we understand it or not, space-time is the fabric out of which the universe is fashioned. And some of the reverberations of this insight may help us to see through the glass darkly. Ta dah! TWIDDERS experiences!

Now let's toy with some of the ramifications of QM.

QUANTUM MECHANICS

When I was a kid, we were taught that the smallest bits of matter were electrons, protons, and neutrons; these three entities made up atoms. We now believe that the smallest bits of matter are quarks; electrons, protons, and neutrons are made up of combinations of *these* particles.

Be that as it may, imagine an atom as it's still generally depicted: a compact nucleus made up of protons and neutrons with electrons orbiting around it.

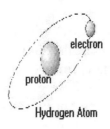

Hydrogen Atom

Electrons spin around the nucleus in shells a GREAT distance from the nucleus. In fact, if the nucleus was the size of a tennis ball, the atom would be the size of the Empire State Building! Atoms are mostly empty space; 94.6% to be exact. (Unless, that is, we get into discussions of quanta-carrying forces that from one perspective neatly fill that space.)

Nevertheless, let's stick with the *heck of a lot of empty space* scenario in an atom. All matter is, of course, made up of atoms.

When I look at my body, I see a much too solid mass. When I sit at my desk, I perceive my desk as another exceedingly solid mass that could be most easily reshaped with a very hard-edged, sharp skill saw but certainly not with the back of my solid-but-soft hand.

Yet, if some guy in the cosmological control room flipped the switch and turned off the forces interacting between the electrons and nuclei of the atoms making up me and my desk, thus shutting down the electromagnetic and/or weak nuclear forces, you know what?

My hand with all of that empty atom space would go right through the desk top with all of its empty atom space like a hot knife through butter. Not to mention my derriere through the chair!

And the significance of that odd bit of minutiae? We're getting to it; hang in there just a little longer!

THE THEORY OF EVERYTHING

The Standard Model of physics does a superb job of describing how sub-atomic particles and forces work with one notable exception: Gravity. In the last few decades, superstring theory has emerged as the most promising candidate for a quantum theory of gravity.

More than just making sense of gravity on a quantum level, superstring theory would really, really like to provide a complete, UNIFIED description of the fundamental structure of our universe in all regards. Yup, it would one day like to be known, with all due respect, as the Theory of Everything!

Superstring theory is based on the premise that all fundamental sub-atomic particles are really just different manifestations of a single object: a string. Rather than such basic units of matter as electrons being zero-dimensional points, they are tiny one-dimensional filaments called strings. Strings that vibrate.

In *Warped Passages: Unraveling the Mysteries of the Universe's Hidden Dimensions* (2005), Harvard physics professor Lisa Randall shares an intriguing feature of superstring theory; it involves the prediction of extra dimensions. At least six.

You think the notion of extra dimensions is just this side of nuts?

Randall explains that the nonstick fry pans in your kitchen are coated with quasicrystals. Quasicrystals exist in multiple dimensions, including one or more beyond the fourth!

What form these dimensions may take is still the subject of hot debate, but exist they do. Evidently.

Well, so much for our speed-of-light trip through relativity and QM, highlighting facets of each that may shed slivers of light on TWIDDERS.

Just before embarking on this fascinating physics trip, I referred to developments focusing on three particular topics that again may help us to understand TWIDDERS: brane worlds, multi-verses, and the Now.

BRANE WORLDS, ET AL.

Imagine 2-D Tom's sheet-of-paper-world blowing on a breeze through our three-dimensional world. Now imagine that our entire universe is much like Tom's domain, just a 3-D sheet of matter—a brane (short for membrane)—moving through a four-dimensional background called "the bulk."

Further imagine that our brane is not the only one wafting through the bulk. Just as more than one sheet of paper can billow about in our reality, so could more than one 3-D brane exist in the bulk.

Maybe these other branes are parallel universes. Astonishingly, scientists believe these universes exist less than one millimeter away from us. In fact, gravity is just a weak signal leaking out of another universe into ours!

And just where might these other brane worlds have come from? Possibly the same Big Bang that got the ball rolling for us. In the article *Inflation and the Accelerating Universe* MIT physicist Alan Guth (2001) suggests that a concept he calls "inflation" might happen over and over in a process of "eternal inflation." Each instance of inflation is the conception of another universe—hence, multiverses. On the grandest scale, he envisions multiverses as a form of interconnected pocket universes.

Or, in *Physics of the Impossible* physicist Michio Kaku (2009), scraps the one-and-only Big Bang and replaces it with continual, ongoing Big Bangs, each one birthing a new universe-in-a-bubble. Our particular neck of the woods represents just one of an infinite number of membranous bubbles, wobbling through an eleventh dimension.

Whatever, however, whyever the distribution of matter, physicist Julian Barbour (Polymathicus, 2007) maintains that when contemplating the nature of existence, we must begin with the premise that there is NO TIME. While Einstein unified space and time into a single entity, he still held on to the concept of time as a measure of change. In Barbour's view there is no invisible river of time. Instead, he thinks that change merely creates an illusion of time with each individual moment existing in its own rightness, completion and wholeness. He calls these moments *Nows*.

Life is moving through a succession of Nows. Barbour explains that each Now is an arrangement of everything in the universe. Once again let's imagine Barbour's *Nows* are like pages of a novel ripped from the book and tossed randomly onto the floor.

Each page is a separate entity. If you picked the pages up and ordered them by page number, it would seem that a story is unfolding. Yet even if you arranged the pages randomly, each page is still complete and independent of all of the others—existing simultaneously. Now.

In this poor shredded novel, we could read on page 4 that Tom, a 2D world hero, had just left on an easterly exploration trip to the edge of his world. Or, we could just as easily snatch up page 374 and read how Tom has ended his journey at the far edge of 2D Land and is contemplating whether to walk all the way back to where he started. (His only choice in doing so would be to walk backwards or upside down—neither option being very appealing to him!) Not past, not future, just Now.

In the great Now of Barbour's universe, all possible configurations of the universe—every possible location of every atom—exist simultaneously. There is no past moment that flows into a future moment because there is NO TIME.

Okay. So what have we got?

Modern physics posits . . .

> Multiple/parallel universes
   Implication: Other realities a whisker away
> Wormholes
   Implication: Free ticket to other "times"/locations
   in our universe or another one
> The illusion of solidity, and manipulation of sub-atomic
   forces
Implication: Travel to other realities with   negligible wear and
tear on our beings
> The great Now Implication: Past, present, future? All
   events are here and Now!
> Dimensions beyond the three in which we routinely
   operate
   Implication: Just because we don't have all of the
   answers, doesn't mean they don't exist!

So, if we do indeed live in just one of multiple universes, and time is an illusion, and in one of these realities Elvis never died, and wormholes assemble and dissemble randomly throughout our existence and can transport us either to an alternate there or other Now, and the atoms that make up our being are perfectly happy to participate in said travel, then might we unwittingly be part players in a moment from history and/or visitors to noncongruent locations? And if we can go there and back again, have others from elsewhere come unwittingly for brief visits to our universe?

Huh?

Consider this. The imprint of a sandal shoe found near Antelope Spring, Utah, in a rock over 500 million years old (Wenlong, 2005). There are even crushed trilobites under foot, suggesting that the wearer walked over them and left the imprint that later became fossilized. (FYI Trilobites, which looked like giant woodlice, died out hundreds of millions of years ago.

Modern humans hit the earthly scene only about 150,000 years ago.) So how the heck???

I've done the best I can to get a grip on the phenomena I've dubbed TWIDDERS. To come clean, I was interested in Mrs. Smith's powder room enigma as more than just an eye-popping account of a really odd occurrence wedged into a book filled with ghost sightings. For years I've been plagued by a seaside experience . . .

WAS IT A TWIDDER?

I grew up in western Oregon. Regular visits to the coast were a family rite. I have many fond memories of sunburnt hours spent searching dunes for the elusive whole sand dollar, the most alien driftwood scraps, and the Holy Grail of West Coast beach combing: a Japanese glass float. But dearest to my heart was the humble agate. Oregon beaches have been so picked over, agates are a rare commodity. I counted it a fruitful foray onto the sand if I found even a pill-sized rock.

As a young wife and mother, my husband and I continued the family tradition. Whenever we could, we'd head to Wheeler—a tiny northern coastal community on a muddy bay—where a kindly uncle owned a vacation home.

One weekend we set out to explore seasides. Not too hard; just head south from Wheeler and take any westerly road. We spent time playing tag with sneaker waves and beachcombing at a few favorite beaches.

Then we headed south again, determined to find one new-to-us beach before lunch. A few miles farther and we turned down a nondescript lane that almost hit the shore and then paralleled the Pacific.

It led to an uninviting little beach, sporting nothing but rock. So much rock you couldn't see sand. A narrow shoulder along a bouldered dike with one set of soggy wooden stairs led us to the stony waterfront.

But oh, the treasure that dull gray rock concealed: agates! I'm not referring to those miniscule mites littering my collection; I'm talking rich, ripe, *plum* sized agates that filled our bucket! Funny thing about that water front street—a dismal row of pealing-paint bungalows faced west like battle weary soldiers. Never an open curtain or a stray dog. Not even seagulls. Just a wicked wind blowing foam off the bleak Pacific, setting our teeth a-chattering in spite of the sunny weather we'd left a few blocks back.

We returned home to Portland with our treasure trove. But the oddest thing occurred the next summer.

We couldn't find that bouldered beach.

How hard could it be?! Go south from Wheeler and take every road to the right. Which we did. Repeatedly, to no avail.

We explored every road from Wheeler to Lincoln City, far beyond where we knew we'd been. We were never able to find that beach again.

I've often thought the simplest, most accurate explanation for why we couldn't find that bouldered beach was that we were idiots.

Maybe TWIDDERS offers another explanation, as it has for so many other folks.

Chapter 2.

Back to the Past
(New World Experiences)

The majority of TWIDDERS experiencers appear to travel back to the past. Why? Not a clue. If every moment of existence is simultaneous, then why do travelers glimpse bygone events on average eighty percent more often than yet-to-be ones?

Shouldn't it be like randomly selecting socks out of a drawer in a dark room? Given that in this particular drawer there is an equal number of black (the past) and white (the future) socks, odds are if we reach in four times, when we get out to the light of the hallway, we'll have two white and two black ones. Shouldn't a wormhole drop us into the future as often as the past?

Is it just that history trips get written up more often than futuristic ones? Or is what has already happened (a set-in-concrete-record merging all probability strands into ONE event track only) a heck of a lot easier for wormholes to transport us to than a gazillion ever-shifting paths of probability event tracks yet to occur?

And if there's only one past and a gazillion futures, then does that mean the sock drawer has only one black sock (the past), and a huge gob of white ones (the future)? So wouldn't that skew the odds towards trips to the future???

Which could lead us into a philosophical discussion on fate versus self-determination. I'd error on the side of the latter and opt for—in spite of the great Now—future visits being, for whatever reason, a lot more dicey.

Okay, the more I chew on it, let's just drop the sock drawer analogy. Sheesh.

Let's begin our time-slip journey on the road most traveled with TWIDDERS to the past. We'll wend our way through old world and new world experiences. Then we'll leave the beaten track and meander down mysterious paths documenting time-slips in more rare forms: fast forward, so slow, instant replay, lost locations, and, oddest of all, alternate reality.

KING SOLOMON'S TEMPLE

Oscar winning actress Shelley Winters appeared in dozens of films, as well as on stage and television. In her book *Shelley: Also Known as Shirley* (1980), Winters shared her TWIDDERS experience. In 1949, she and then-fiance, Farley Granger, waylaid a whirlwind European tour for a quick trip to the fledgling state of Israel.

Raised in a Jewish family, Ms. Winters' feelings, returning after centuries of the Diaspora, were very tender. Much to the pleasure of Israeli authorities, Winters and Granger were the first American movie stars to come as tourists.

Winters spent the next few days in Tel Aviv where, she wrote, "Nearly everyone was trilingual at least . . . the only jarring thing was that since everyone had summery short-sleeved clothes on, every now and then I would see an inner arm with a number tattooed on it" (pp. 337-339). The marks served as enduring reminders to the horrors of Hitler and the Holocaust.

On their last day in Israel, the celebrity tourists scrunched in quick trips to Jerusalem, Nazareth, and King Solomon's Temple.

The sights and sounds of Jerusalem were such a profound mystical experience that I, who do not believe in mysticism, cannot even begin to describe it. We wandered around the famous sights, and I never opened my mouth . . . I, who never shut up.

It was dusk when we got to the ruins of King Solomon's

Temple . . . and we seemed to be the only people there. As everyone was turning around to go back to the car, I leaned down and picked up a very, very small stone. I knew that in years to come, if every tourist did that, pretty soon there wouldn't be any ruins left, but my spirit was sorely in need of Solomon's wisdom.

When I looked up, there was a man smiling at me. He was about 20 feet away and had on an old reddish brown robe; he had longish hair and was leaning on a staff. I thought he was the caretaker [in costume] so I guiltily dropped the stone . . . he pointed to the stone and gestured for me to take it. I picked it up again.

[I] ran back to my companions and asked . . . 'is that man the caretaker of the temple?'

We all looked back to where I had been standing, but there was no one there (pp. 343-344)!

THE FARMHOUSE

Terri Leigh Moore (1999) recounts her mother's TWIDDER encounter. Moore's mother spent her early years at a place called Devil's Lake, which has since been renamed Good Spirit Lake. Terri's grandma often sent her then-ten-year-old mom on errands to the nearest store. The store was about a half-hour walk away, and walking was the mode of transportation for this youngster.

One morning her mom was walking to the store for a few items when she passed a farm. There were a dog and a bunch of chickens out in front of the house. A little girl was playing in the front yard. She appeared to be about ten. Terri's mom and the farm girl started up a conversation. Her mom decided to postpone errands for a while and stay and play with her newfound friend.

Lunchtime rolled around, and the little girl's parents invited Terri's mom to stay for lunch. "My mom totally remembers the lunch because she came from a poor family and the meal was so extravagant for her," Terry recalls. The family was very friendly, and Terri's mom was having a grand time.

After lunch, she played for a bit more but realized it was getting late and that she'd better get to the store and back home. She ran the rest of the way to the store and back in hopes of squeezing in a little more playtime with her new friend before finally getting back home.

"Well," says Terri, "she gets back to the farm yard, and there is nothing there. Not even a chicken feather. The house, which had been this beautiful old house, now looked like it was falling apart and was all burnt." Worried now about getting home, she didn't spend a lot of time trying to work out the "how the heck" puzzle of her new friend.

When she at long last arrived home, she got the "hugest lecture as she'd been gone so long and everyone had been worried about her."

She explained about being waylaid at this house and playing with the nicest little girl. Terri's grandma scolded her daughter, accusing her of lying. Her grandma told her in no uncertain terms that no one had lived in that house for over twenty years—since the family who lived there had died in a huge house fire.

Terri relates that her mom never saw the little girl or her family again.

GETTYSBURG

In 1981 then-13-year-old Christopher and his family passed through Gettysburg, Pennsylvania, while on vacation. He wrote about his experiences there.

It was summer and the trees in all of their full-leaf glory were like jewels on the beautiful Pennsylvania landscape.

It was a hot day with the sun shining and the blue sky was dotted only occasionally by a passing cloud. It was the kind of day that would have invited even the most melancholy heart to bask in cheer and tranquility.

We visited all the significant residences, battlefields and museums like any respectful tourist, making sure to read each and every informative marker.

During the day, it was mentioned how quiet I had been and by the end of the day, the observance was again duly noted and embellished. Of course, at that age (13) I was known for the inability to control the growing muscles in my mouth and the ever-increasing opinions flooding my maturing mind. So, for me to remain quiet most of the day was almost of as much historical significance as the Gettysburg Address!

One would assume, as did my family, because all of the sites and fascinating information occupying my thoughts that, like Microsoft Windows trying to multitask, I was rendered speechless. In all actuality it was something far deeper and more sobering than that. It took me several years before I was willing to reveal what had happened and only then to a select few.

We were standing on one of the battlefields looking across the grassy expanse. Suddenly I was gripped with deep sorrow completely in defiance of the beautiful day. The sun disappeared into insignificance and the horror and darkness of bloodshed hypnotized my attention. Everywhere I looked I could see bodies—bloody bodies of soldiers who lay dead or dying.

I looked at each body and the one closest to me was a man in his mid 30's with dark hair and a full beard that traced his jaw. He lay on his stomach with his face toward me, his eyes closed forever. Yet, his soul was so real I felt like he would look up at me any second. He was a Confederate soldier as I can remember vividly his gray uniform. His canteen had not fallen far from him, and I could see his knife and another leather pouch of some sort lying on top of the tall, damp grass.

As I stared in disbelief, I heard cannon firing in the distance and listened to the cries of men in agony. The smell of blood and mustiness permeated the heavy air, and I had the unexplainable feeling of wanting to die.

Just as quickly as the phantom scene appeared, it vanished leaving the moment embedded in my memory with every detail intact. Images are one thing, but when they are accompanied by emotions—emotions of sorrow, despair and pain—they become experiences that chill an observer like an arctic breeze. Even now, with every recollection of that phenomenon, I am silenced with soberness.

What I saw was a scene . . . with three-dimensional sensations of smell, taste, and presence.

SHARPSBURG

The Battle of Sharpsburg (Antietam) was the bloodiest single-day battle in American history with about 23,000 casualties (OPFOR1, 1997). It took place near Sharpsburg, Maryland, and Antietam Creek.

One young man and his three friends were cruising when they decided to tour the battlefield at night. They thought it would be interesting "simply because it was a beautifully preserved historic battlefield with the mountains in the background, and quaint town behind us."

Anyway, it was about 2:00 or 3:00 A.M., and we were going down the road passing the north woods headed towards the cornfield. It was at this cornfield that troops under General Jackson (CSA) was attacked by General Hooker (USA) and pushed back.

General Hood (CSA) counter attacked and pushed Hooker back, and thus did the early part of the day go. By the end of the day, the corn, which had been higher than a man, was cut lower than boot heels due to gun fire and bayonet charges.

While traveling down the road at approximately 30 miles per hour, we saw a line of men crossing the road about 200 yards away from us. Not wishing to hit them or frighten them, I slowed down. As I was slowing down, my friend Jason made me aware that those people were dressed in Civil War uniforms.

As I got closer I could also see they were in Civil War kit. I really thought nothing about it since re-enactors do reenactments all of the time. The thing was that as I got to within 50-75 yards of them, they disappeared before us. I mean the people on the road just faded out.

Stopping the car, we got out and looked about a bit, but neither saw nor heard anything. There must have been at least 100 people moving across the road at that time, and you just can't hide the sound of that many people so quickly.

We were never scared but more confused since we used to go up there quite often and never saw anything like that before in our lives.

When we got back to my home, I looked up the Battle of Sharpsburg and found out that it was fought on September 17th, 1862. The night we saw this was September 17th, 1992!

THE CHURCH

In the online story "The Vanishing Church," Patricia Tallberg (2002) of Kansas recounts a TWIDDER from her junior year in high school.

One spring day some friends and I decided to go driving on the country roads outside of our town. We did this for about an hour until we came upon an old church with a graveyard beside it. Brian, who was driving, decided to stop and investigate.

We all climbed out of the car and started to look around. It was a warm day but for some reason when entering the churchyard it felt as if the temperature had dropped about ten degrees. Nancy went back to the car and put on her jacket, saying she was cold.

The graveyard seemed to be well kept though there were no flowers on the gravesites. This seemed strange to me since Memorial Day was that weekend. There were no gravestones dated any later than 1931 even though there was plenty of room, which also seemed a bit unusual.

Brian wanted to go into the church which was boarded up, and having nothing better to do, we all agreed. We thought that since the windows were boarded up, that the door would be barred or locked. We got ready to have to 'break in' when Brian pushed on the door and it swung open. At this point we all became a bit nervous but went inside.

I remember the smell hitting me first. Instead of being old and musty, it smelled like flowers—roses, to be exact. The place was spotless—not a speck of dust anywhere. We looked around and couldn't figure out why if this was an abandoned church, and believe me, from the outside there was no question about it—the inside was so nice and clean?

The next thing we noticed was all the Bibles sitting in the seats, as if waiting for the congregation to sit down and pick them up. We all looked at each other and I said, 'I don't think we should be here.'

My friends nervously laughed, and Nancy said she needed to get home and besides she was getting hot. Brian and Mark, being teenage boys, said we were just being sissies and told Nancy to take off her coat if she was hot because they weren't done exploring.

Nancy and I drew up our courage, and I figured it was just nerves. As we stayed longer, I noticed that dust started appearing—not a little bit but thick coats of it. It was as if the inside of the church was aging rapidly to catch up to its outside appearance. Brian and I watched as a spider web just appeared between one of the pews, and at this point all of us got a very bad feeling. We all ran out of that church as if the devil were on our tails!

As we drove back, Nancy remembered her jacket. Brian told her not to worry; we would go back tomorrow afternoon, but he was not going there today because it was getting dark. I could tell it was more to do with the feeling we got from the place rather than being late getting home.

We wrote down the road numbers at the closest intersection to the church and put down the directions to the nearest highway.

The next day was Saturday, and when I got off work, we all climbed into Brian's car and followed the directions back to where the church was. What we found was a dead end and a local lake in the vicinity of where the church should have been. Brian said we must have written down the directions wrong, so we drove around for another hour and still no church. I finally said we should go to the nearest town, which was only about five miles away, and get directions; being tired and frustrated, Brian agreed.

We pulled into a convenience store, and there was an old man working there. I thought if anyone might know where our mysterious church was, he would. When we described the church and graveyard and asked where it was, he got as white as a ghost.

Brian started repeating what we had said and the old man stopped him. He said, 'There is no way you kids could have been to that church. It burned down in 1932 when some drifters tried to light a cook fire inside; then the state put in the lake in 1935. That land where the old church and graveyard sat is under 50 feet of water.

We thought maybe he was talking about a different church, but when he said his little sister had been buried there and said her name, I knew it was the same place. Just before we went to go inside the church, I remembered looking at her tombstone and thinking how sad that a little girl so young had to die.

I don't know how or why the church appeared to us that day, and I don't know what would have happened to us if we had stayed there much longer. To this day I am grateful that we left when we did, or I might not be here this day to tell the tale of the vanishing church.

THE BRAVE AND THE BUFFALO

In the summer of 1985 Keith Manies was doing traffic studies of remote rural highway intersections for the Kansas Department of Transportation. He said it was a dull job, but it afforded him time to read and earn money for college. Manies (2008) told his TWIDDER story that took place on a particularly hot July afternoon.

I found myself doing a traffic study overlooking the Smokey Hill River Valley in west central Kansas. It had been a rather uneventful day. All of a sudden I heard a very high-pitched noise akin to electronic feedback. My first reaction was to turn down the car radio, which I did—to no avail. The irritating sound seemed to come from the back of the vehicle, so I got out to investigate.

As I walked down the road, a movement to my right caught my eye. I turned my head and to my utter amazement, there, coming down the highway embankment was an Indian on horse! Now, this was not any modern-day Native American, but a real-life snapshot of an Indian brave circa 1840. He had a leather loincloth and a pair of moccasins. He was riding bareback without any conventional bridal, just a rope tied around the pony's head. In the Indian's right hand was an antique-looking rifle and in the other a rope.

The sight took my breath away and all I could think of was to say 'Hi!' The Indian did not respond to my excited greeting and totally ignored me as if I was not there. As the Indian approached the far side of the road, he stopped and intently scanned the river valley below.

The Indian's stare made me turn to see what he was looking at. Down in the valley, again to my amazement, was a large herd of buffalo strung out for miles where there were none two minutes before. This sight made my head swim because the mighty herds of buffalo had been exterminated from Kansas over 100 years previous.

I caught my breath and turned back to observe the Indian and his horse, but there was nothing there! I ran over to where I had last seen the Indian and looked down the hill. There was no sign of the Indian and now no trace of the buffalo either. As the hot sun beat down on me, I slowly walked back to the car and noticed that the annoying sound was absent, too.

As I sat back down in the front seat of the car feeling somewhat light-headed, I pondered what I had just witnessed . . . had I experienced a hole in time?

THE FIELD WORKERS

Brandie (2005) shared her TWIDDER experience that occurred on a summer day. One afternoon as Brandie was coming home from work, she crested a hill with a watermelon field on the left. She noticed several men.

They were sitting around eating watermelons in the field. About 20 or so melons were busted up and scattered all over the ground.

As I approached them, the men all looked toward my car. One man looked me right in the eye and started running right in my path. I slammed on the brakes as hard as I could. The man kept going, and I missed hitting him by less than a foot.

I was so angry that the guy would pull a stunt like this that I turned and looked at him. He was on my passenger side just at the rear of the car. I flipped him off while expressing my anger pretty loudly through the open windows. I could hear all of his friends laughing at me.

Then I noticed that they were all dressed in funny—like clothing from the 1920's with hats on, and some had scarves tied around their necks.

I drove on home, which took about two minutes, and went in and told my dad what had happened. He said he'd go down there and say something to them. He was gone only a few minutes.

When he got back, he told me that I'd better quit making stuff up and to be more careful driving! He told me there were no men there, and he could see my tire marks where I'd slammed on the brakes.

He thought I'd made the story up. I told him he must have looked at the wrong place and that I'd go back with him and show him.

This was less than 15 minutes after I'd first gone by the field. When we got there, the men were gone, and all of the watermelons were intact.

We got out and looked. There were no footprints in the field at all, and there should have been because the ground was all muddy. That is the weirdest thing that has ever happened to me.

THE FORT

Mr. Torrence (2004) told his tale of slipping back in time. For Mr. Torrence, it started back in 1999 when he was 19. "I had an experience that I can't explain that freaks me out to this day."

My girlfriend and I went up to Poughkeepsie, New York, on a trip to visit her sick uncle, a man nicknamed Floyd (Mayberry reference) due to his profession as a barber although he was also an avid biker. He had lupus and about a thousand other ailments, it seemed.

He lived on a large tract of land that included nature trails that seemed to go back for miles on up into the woods. One morning after breakfast, Cindi (my ex now) and I decided after breakfast to head out to a location that she remembered going to when she was a kid. She called it the Big Indian Rock, which she said looked like a profile of an Indian chief.

So at around 10:30 or 11:00 or so, we headed up there. It was a long way up into the woods, but Cindi, after getting some directions, seemed really confident in spite of not being there in years.

After a while, I was almost sure she was lost. But she saw something that she recognized and headed off in that direction. It was this large white rock with a fort into the side of it like a kids' fort or something.

Her exact words were, 'Oh my gosh, I remember this

place!' and then she began to tell me how she and her friends, a bunch of boys who lived nearby, had built this one afternoon. She was totally surprised to see it after all of this time.

This was at least 10 years later, and this poorly constructed fort was still there, which I thought fascinating, too, at the time. The fort consisted of a low lean-to, almost like a box made of rotted plywood scraps.

It sat to the side of the huge white rock and had a bundle of dirty camping gear inside of it, sleeping bags and other dirty junk and leaves, cobwebs, etc. Cindi went up to the entrance of the low-to-the-ground fort and looked in at it as I jumped up on the top of the white rock to have a seat.

I looked over to where Cindi was, and she had disappeared from view. At first I thought she maybe crawled inside, but on more careful observation, this wasn't the case at all. I turned to see whether perhaps she had gone around the other side of the huge rock, and when I turned, I saw two little freckle-faced boys, one taller than the other, staring at me. One wore a ball cap on backward and the other was in a dirty flannel jacket. They just stared at me eerily and then suddenly it just seemed like time slipped and suddenly I was walking up a wooded path towards a clearing, and there was Cindi telling me to hurry up and come on.

My first question to her was who were those two boys, and she said, 'What are you talking about? What boys?'

I said, "At the rock."

She just punched me in the arm and began to point out the face in the rocks. She told me how she used to be a tomboy and used to ride BMX bikes back there and other such memory lane kinds of stuff.

After a few minutes we headed back to Floyd's place, and we were greeted by Cindi's mother, who'd just arrived.

After a while, Cindi came in and we all got to talking and the fort came up. She asked about some kids she knew in the area and if they were still around. Floyd got real dark when Cindi mentioned the Moriarty kids and got quiet.

He said to Cindi, 'The two boys and their mother were killed by their father before he took his own life." (Having something to do with a divorce.)

When I heard this, the hair on the nape of my neck stood up. While driving back, Cindi kept the two boys as a topic, and said that they both had a crush on her. And once even fought over her, and man, that just made me drive faster! We made record time back to Jersey.

THE TRAVEL AGENCY
    Cheri Owens (2004), a good, kind friend, went on an errand for an acquaintance and her children, and ended up having a startling experience.

I'd agreed to purchase airline tickets for my friend's two kids by a certain date. I was out shopping on the last possible day that I could get them when I realized that I'd completely spaced the tickets off.

It was 4:30 P.M., and there was no way that I could make it to the airport to get the tickets. So I stopped at a phone booth and looked in the yellow pages for travel agents.

There was a listing for 'Macy's Travel Agency,' and the address was directly across the street from where I was. I called the number in the phone book and a man answered, 'Macy's Travel Agency.'

I asked him what time they closed. He told me that they closed at 5:00. So I asked him if they were located inside the actual Macy's store or just in the plaza itself. He gave me these exact directions: 'Go inside Macy's store. Go to the center of the store. You'll see the escalator. Go up to the third floor. As soon as you step off the escalator, if you look straight ahead, you'll see a small office area that's all lit up. Go to the desk, turn right down a small hallway, and we're the second door on the right.'

I hung up the phone and ran across the street. I got on the escalator going up, but as I was nearing the third floor, I realized that there were no lights on up there. As my head cleared the floor, I looked around, and there was nothing up there except tools, boards, and signs that said, 'Please excuse our dust.' I decided that as soon as I stepped off, I would just run around to the other side and get on the down escalator. But as I stepped off, I noticed that there was a small office area that was all lit up directly in front of me. I mustered up all of my courage and I walked to the main desk.

I could see the hallway to my right, but I had forgotten which door I was supposed to go to. I rang the bell and a young woman came out of the back. I told her that I couldn't remember which door was the travel agency. She looked confused and told me that she didn't think that there was a travel agency in the store.

I was too impatient to argue with her, so I asked to see her supervisor. A man came out and I asked him the same question. He shook his head and told me that he thought there might have been a travel agency there a few years ago, but it had been gone for at least three years. I told him that I had just called the travel agency's number a few minutes ago, and the man had given me directions to get to this office on the third floor.

He asked me where I had gotten the phone number to call. I told him it was in the phone book. He came around to my side of the desk and pointed to a payphone located on the wall of the office area. He handed me the phone book and told me to find it. I looked it up in the Yellow Pages and showed him the number. He handed me a quarter and told me to call the number. So I did.

It rang once and then I heard a loud tone, and a recording said, 'I'm sorry, but the number you have dialed is either disconnected or no longer in service . . . .' I was holding the phone so that he could hear it also.

I turned and looked at him. He smiled at me and said, 'Do, do, do, do, . . . ' (the theme tune from *The Twilight Zone*). I dropped the receiver and ran to the escalator as fast as I could. I've tried to figure it out, but I can't. I've never shopped at Macy's since.

JURASSIC PARK?

Y. Phillips (2009), of the grand ole South, went on a hunting trip with his granddad in July, 2008. He wrote about his unusual experience.

I don't see my grandpa very often, so I always take the chance to take trips with him. Grandpa is pretty much an

36

outdoorsman and enjoys hunting, fishing and just being out in nature.

Grandpa and I were out in the woods. It was around 3:00 to 3:30 on Friday the 25th of July. I was 18 at that time. We were on Grandpa's land in Georgia.

It's a pretty place with the typical Georgia woodland and a few grassy plains. We were walking on a little rocky road heading for a site where Grandpa often sees deer. As normal, there were a lot of sounds going on at night in the woods. We ignored most of them and remained quiet to not scare away anything.

Suddenly we heard an unusual noise we never heard before on our many hunting trips. Grandpa looked at me and listened. Then he raised his finger in front of his mouth to show me that we shouldn't make any movements. I heard a lot of movement and more of the noise. I can't really describe the sounds, but I sure can describe what I saw, even when it was pretty dark.

We just kept listening to the sounds as suddenly something came walking slowly out of the bushes and onto the road maybe 150 yards in front of us. My eyes got really big, and at that moment I wasn't even scared, just amazed to see this creature. We didn't move. As crazy as it sounds, it looked just like a raptor from the popular Jurassic Park movies.

I just froze because I thought things like that lived many thousands of years ago. It had a long, stiff tail, walked on two feet, and had short arms. It looked lizard-like and had a huge claw on both of his feet and smaller claws on his arms.

Since the creature appeared to us that it could run fast, we decided to just not move at all. It raised its head in the air, and it seemed like it was smelling the air. I estimate its height around 5 feet at the shoulders. After sniffing the air, it made these sounds again and turned around and ran off in the bushes.

Grandpa and I waited until we felt safe again and then quietly made our way back to the truck and drove home. In the truck, we talked to each other about what we had seen and decided to not tell it to Grandma because she would think we were crazy.

I never believed in stuff like ghosts and creatures and paranormal stuff, and I still don't believe in ghosts. But since that encounter, I believe in creatures that science doesn't know about. That's my story, as odd as it sounds. I know what I saw.

JURASSIC PARK II?

James (2002) wrote about an experience he had in winter, 1998, while employed driving a tractor for a mining outfit.

I was driving a tractor on a dirt road overlooking a 200 foot mining pit nine miles southwest of Corona, California. It was a mining site at the foot of the Cleveland National Forest—a somewhat remote area.

It was very quiet except for the diesel engine of the tractor. I sensed a presence above me, and looking up, saw what seemed to be a very large, leather-winged, sharp-beaked Pterosaur. No feathers, bony wings and a somewhat long, narrow tail.

It seemed to have a bit of red on its head and dark, dull

green on its back as it passed away from me—probably 20 feet above me with a wingspan of 30 feet.

Its flight was slow as it flapped its wings and seemed to glide rather than fly like a regular bird. It didn't notice me at all or the noise of the diesel engine of the tractor (which was just fine with me).

All around this Pterosaur, the air appeared to become wavy, almost like the effect of ripples in still water. I don't know why, but I got the feeling that it was in another time zone or just didn't belong in this dimension.

As I sat watching on the tractor, the further it flew away, the more it looked like it was dissolving away rather than growing smaller in the distance—and still with air waves around it.

There are rare California buzzards around, and I have seen them in person. Though they can be quite large, this being dwarfed any buzzard. My opinion is that this flying creature was something that should not be alive at this present time, but nevertheless I watched it fly over me and away for at least 100 yards.

ANOTHER CHURCH, ANOTHER SLIP

Another TWIDDER experiencer, a young mother whose web moniker is Yknot1 (April, 1995), wrote about her experience.

I was almost seven and on vacation with my parents in Quebec, a city I had never been in before. I remember walking behind them and thinking how familiar everything looked. My mother turned into a store, and my father and I both turned to follow her.

But for some reason, I didn't go with them (they thought I was right behind them) but wandered into a church that was just down the street.

The church was smoky and the people dressed so strangely. There was some sort of service going on, and I sat down in a pew and listened.

I remember thinking how funny the people were dressed when I suddenly realized that instead of the clothes I had been wearing, I had on a long dark skirt. It scared me so that I got up and ran out of the church and right into my parents, who were looking for me.

I remember telling my mother that I was scared, but she only looked at me funny. Then I started crying and told them that they were doing something strange in the church. This time they understood and took me back into the church to see.

It was entirely different . . . not smoky, just a few people in normal clothes . . . I continued to cry and my parents took me back to the hotel.

We never spoke about it until years later. I almost forgot about it, and yet at times it would come back to me. Then after I had lost my daughter for a few minutes one time, I was telling my mother how scared I was. She asked me if I remembered being lost in Quebec.

And then she said that she never understood where I had learned French so quickly while I was there. I had no idea what she was talking about, but she told me that when I ran out of the church, the first words I said had been in French!

HAWAIIAN GRAVEYARD

A fellow who also prefers anonymity (8lck89$2kq0$1, 1995) shared his experience that occurred on the Big Island in 1980.

I was very interested in photography and went to take pictures of one of the large stone temples on the Hilo tip of the island. This temple was near the ocean on the big plains of rubble-type lava.

It had a sign saying that it was sacred to the Hawaiian people and please don't trespass. Being a firm believer in bad karma, I didn't climb up. I contented myself with shooting about half a roll of film at about 1:00 A.M. by the light of a full moon. I wasn't looking for anything ghostly, I just wanted some really neat shots of a pretty impressive temple.

Having finished shooting, I went back to the car. I noticed that on the other side of the road, the ground dipped down about six feet where there was a very large area that was clear of underbrush. There were a few trees and it was very brightly lit by the moon.

I climbed down. I thought at first that it might be a picnic area since the sandy ground was clear of plants and looked like it had seen a fair amount of foot traffic that day.

As soon as I started walking around, I realized that I was in a graveyard. I wandered around, avoiding graves (no headstones; they were just very obviously burial places).

I went and looked at a small stream that went down to the ocean here (more undergrowth there, looked like a good place to get bit by something).

I walked inland a little bit, the place got more ravine-like, more undergrowth, less light—so I went back to the graveyard.

I walked a few yards to where the ocean came into a very small cove and admired the moonlight on the waves for a while, drove back to where I was staying, and went to sleep satisfied that I had some interesting pictures.

Later on that trip, I had an opportunity to pass that temple in the day time. I stopped to get some daylight shots of the place and then crossed the road to get some day shots of the graveyard.

I was vastly surprised to find no graveyard. The ravine was there. The ocean and cove were just like I remembered, but the floor of the ravine was full of underbrush and densely wooded. There was NO open area and nothing resembling burial sites! I was very surprised.

I climbed down and gave the place a through exploring (even climbing up the ravine to where it was crossed by the road). I found some neat stuff, including a very shallow cave with some petroglyph-style figures and old woven reed mats but no clearing and no graveyard. The small stream was there, and I followed it from ocean to highway. I found disturbed areas of sand where I had been walking during my nocturnal visit.

I know it's an odd story, but it's true.

MOUNT LOWE

In her book, *Time Storms*, Jenny Randles (2002) told Swede Bo Orsjo's story. He had moved to the U. S. West Coast in 1974 and took a hike alone up Mount Lowe, just outside of Pasadena, California.

He was impressed by the quaint green hotel he encountered halfway up the mountain. He sat at its edge, eating his packed lunch, watching a maid sweeping up in the strange *misty* light.

The problem was that nobody believed him because they insisted there was no such hotel up this mountain. In the end he returned with a friend, determined to show him, but they found only rubble.

Research quickly disclosed that a millionaire called Lowe had once started to build a railway but only got halfway up the mountain before his money ran out. Instead, he constructed a magnificent hotel on the railroad to nowhere.

Orsjo confirmed that the photographs show the same place he visited. Unfortunately, that was not possible because it burned down in 1937, and by the time of his mountain hike it had long since crumbled into ruins.

THE TENNIS COURTS

In August 1997, a couple of guys had a most unusual experience. Nick V. (2000) wrote about a challenging experience looking for a place to play tennis.

My friend Larry and I got together in Cambridge, Massachusetts, around noon to play some lunch-time tennis. We'd never played in this particular part of town before, so we drove around looking for some courts.

I drove up to a policeman who was directing traffic around a small construction site on a side street and asked him where the nearest courts were. He thought for a moment then gave us clear instructions to courts only about two blocks away.

Following his instructions, I took a right into a driveway. Sure enough, right in front of us were the tennis courts. All three courts were occupied.

It struck us both as odd that everyone was dressed in white. We noticed a very attractive young lady wearing a white tennis dress in the far right court, as she was preparing to address the serve.

Still looking at the courts, we decided to wait for one to come free. I pulled my car into a small parking lot directly on my right hand side. When we got out of the car, everything we had been looking at was gone!

The tennis courts were no longer there. A small field on the left was gone. Instead there was a cement building that we did not see when we pulled in. It was as if we were somewhere else.

Strangely, neither of us was particularly baffled at that moment. I felt very annoyed for some reason. We both agreed that what we had just observed was 'too weird,' and we went off to play tennis somewhere else. It was only while thinking about it later that we spoke about it.

Afterward I did some research into the area. Indeed, at one time there were tennis courts there, but they were torn down in 1954! Did we travel back in time? We have no doubt about what we experienced. I sometimes wonder, '*Did the tennis players also see us?*' I may never have an answer.

THE CASE OF THE MISSING RUMMAGE SALE

A gal with the moniker Traveler (May 2002) shared her TWIDDER experience about a time that she absolutely, positively could not find a rummage sale.

The thing that happened to me, which I have dubbed 'The Case of the Missing Rummage Sale,' doesn't sound like missing time. It sounds more like people and an event that are missing. Anyway, here is what happened to me.

My boyfriend drove me to an outdoor rummage sale. It was held in the parking lot of the recreation building of the Sun City Anthem housing development in Henderson, Nevada.

The sale was to start at 7:00 A.M. on April 21, 2007. We got there at 7:30 A.M. on April 21, 2007. There were just three empty parked cars in the parking lot. There were no people, no tables set up, no rummage sale signs . . . nothing. We drove around the parking lot for 25 minutes and then we left.

I've been going to this sale every six months for several years. There's only one way in to this place, only one community parking lot into the entire complex, and only one recreation building.

On Monday, two days after the date of the sale, I called their recreation center to see if they had rescheduled the sale for a different date. The event coordinator told me the sale went on as usual on the date I was there in the parking lot where I was. She said the parking lot was packed with shoppers at 6:30 A.M., and she doesn't know why I didn't see anyone!

There is only one recreation building, and I was there. Tuesday night, I went on the website of this complex to check out the rummage sale. It turns out that I was there at the right place, right date, right time.

45

Where was everybody? Why didn't I see anybody? I've talked to a few people, most of whom definitively say, 'You went to the wrong place, you went on the wrong date, or you went at the wrong time,' all of which is nonsense. I'm slowly calming down after this strange incident, but I will never forget it.

MYSTERY CABOOSE

Jermaine (January 2008) shared a story from when he was around 13 years old and living in Apex, North Carolina. A railroad track ran right in front of his house. It was here that he experienced a TWIDDER.

My friends and I had never seen the track used. I never asked my mom if it had been used, and no one has ever talked about a train going down those tracks. So . . .

We used to play and walk on the track all of the time. One day I was going to cross the street to play at the neighborhood park, and to do so I had to cross the tracks. As I was going to cross the tracks, I saw that there was a red caboose coming down the tracks towards me.

I thought to myself, 'Wow! A train! I've got to go get my brother.' So I turned and ran to my house and told my brother, 'There's a caboose coming down the tracks!' So we looked out our window until we saw it just around the curve coming from behind the trees. We ran out there and stopped beside the tracks about 50 feet away.

We saw two men in the caboose. One was in the front and the other was in the back of the caboose. We waved at the men, and they waved back at us with big smiles on their faces.

We just kept looking and waving, and they kept on going. We stayed there and watched them go down the tracks. When it was a ways down the tracks, we said, 'Let's go and tell the other kids what we saw.' We went across the street and told about five or six others, and we waited at a store in front of the tracks for the train to come back up, but it never did.

We decided to walk down the tracks and see it. We walked all the way to the end of the tracks—about three miles—and there was no turn off for the tracks; the track ended and was covered with rocks and dirt, like a driveway was built over it.

When my brother and I talked about what we had seen, we both said that we saw two men in a red caboose, one in the front and one in the back. Both men had very pale skin and were wearing light blue-and-white striped coveralls and big hats.

Years later, I asked my aunt what kind of suits the train track workers wore back in the day, and she said light blue-and-white stripes with big hats. She said this before I told her what I had seen.

THE BAR SCENE
Donna Bailey (n.d.) wrote about a visit she made to her dad's home in the "Mother Lode Gold" country of California.

I was staying with my father and stepmother Lois while she was recovering from an illness. Lois' room was off the downstairs hall in the back. Dad's was across from hers, and I was in the guest room next to the family room. The second story had been empty for years.

47

It was winter, and I was sitting in the old dining room which is now called the family room. It was about 11:00 P.M., and I was watching television. Lois and my father were both asleep when I heard her trying to get something from her night table.

Sounds echo through big old houses, and I could hear very clearly the small bottles rattling, and the unmistakable tink, tink, rattle, clink. I could imagine her leaning over her potty chair at her bedside, reaching for a Kleenex or water glass.

I'd gone in there quite a few times giving her this or that during the evenings before. The rattlings continued until I heard a loud clunk. I knew something was going to break, or she was going to fall off the bed into the john.

I pulled myself out of my comfortable chair with a small bit of irritation and walked quickly down the long darkened hallway. In one stride I swung her partially open door wide, stepped in, leaned over the bed and said, 'What's the matter, Lois?'

Six rough-looking men stood with their backs to a long heavy bar with glasses raised at the side of her bed. A huge gaudy gilded mirror hung behind the long dark bar, reflecting a chandelier we didn't have. They looked like miners or drovers of some kind. They didn't make a sound but looked at me as wild-eyed as I was looking at them!

Lois was on her side facing away from the men who stood beside her bed. Her water glass and vials on her night table blended with and faded among the dark clothing and long legs of the men.

The men shifted their weight. One man raised both arms straight up still holding his shot glass as though he was at gunpoint! Another, a small stout fellow with a rather bushy moustache grabbed his shirtfront with both hands and dropped his drink.

Then suddenly, freeze frame, the entire scene stopped all movement. No sound came, nothing. They all continued to stand with their glasses raised with surprised looks on their faces as I slowly began to turn around. I swung around scrambling wildly for the other side of the door with my other hand, caught the knob and like in a dream slowly turned, ever so slowly reached the doorway, and even slower yet went through it.

I was pulling myself with every ounce of strength I had. My feet were leaden; my step began to stagger by the time I reached the lamplight. I slammed my back into the chair and listened. My heart was pounding, and I was gasping for air so loudly it was hard to hear.

It was deathly quiet as I sat there listening for sounds from down the dark hallway. Soon the sounds came again. Small tinklings, rustlings and then the unmistakable clatter of a bar, like it was perhaps a few doors away. Bottles on glasses, clattering, clinking and the dull whack of a full bottle on a heavy old bar, the knock of a heavy shot glass. I could hear many muffled voices like a room full of people you hear through a closed door, but you can't quite understand the conversation.

I didn't go back [into] Lois' room that night. The noises stopped a little while later and didn't return for quite a long while. When they did return, they were more faint. I listened to them, but I stayed in the chair. I didn't tell anyone because I wanted to think about every detail with no outside influence. Besides, I wasn't sure anyone else could hear them but me.

A few weeks later my sister came to visit, and on this particular night again came the clinking and clattering from Lois' room once more. I ignored it until my sister said, 'Aren't you going to go in there? She's going to upset the whole nightstand.'

I said, 'If you want to see something good, you go.'

She looked a little puzzled but got up and went to Lois' room. She came right back out and said, 'She's turned the other way. She wasn't doing anything! I don't know what it was; what was I supposed to see?' she asked.

I was a little disappointed to say the least, but at least someone else could hear them besides me. From the relatively calm look on her face, I knew my sister hadn't seen the boys at the bar as I had, but then she added, 'What in the world does Lois have that long, heavy table in there for?'

I've heard the 'boys at the bar' many times since then, but I've left them alone.

THE TRAIN STATION
"UnBreakable" (2009) wrote about a time slip on his morning train commute.

Okay, let me first preface this by saying that if somebody told me this happened to them, I wouldn't believe it. But this is my off-the-wall experience.

I have been commuting to work by train for twenty-five-plus years now, and this occurred about ten years ago (1999).

I arrived at the train station near my place of employment as usual about 8:00 A.M. like every other morning. I was walking through the station with hundreds of other commuters with the usual hustle and bustle.

Suddenly, without warning, it was like I entered another room. The air got thick and hazy and all of a sudden it was difficult to walk, like walking in mud.

People on both sides and in front of me appeared to be dressed in attire from the early 1900's; some men in bowler hats and women in long period dresses. The interior of the station, which I know like the back of my hand, seemed different. There were unfamiliar storefronts where the usual coffee and newsstand kiosks would be.

As quick as it happened (it was like stepping out of that room), and everything went back to normal. I cautiously looked behind me and the usual multitude of people were walking behind me as normal. This only seemed to affect me.

Although I have an open mind, I lean toward skepticism when it comes to time warps, interdimensional travel, etc. However, I have no other explanation as to what happened to me that morning. I wasn't daydreaming, hung over, or anything else. I've had no experiences like that before or after.

Chapter 3.

Back to the Past
(Old World Experiences)

In the previous chapter, there were a couple accounts of prehistoric creature sightings. Quick question: did the observers truly travel to the past, or did the critters travel to the present? And what about Keith Manies, the Kansas Department of Transportation worker who saw the brave on horseback? As an after note in his communication, Manies stated that he was not at all sure that the Native American and mount had not come to his present from their past—rather than he traveling back.

Or Shelley Winters and her interaction with a sabra-from-the-past? Did she travel back or he forward? She never mentioned the road or cars, or her friends temporarily disappearing, or the temple changing in appearance ... so was *the sabra* the one experiencing the TWIDDER and she simply a surprised observer?

In many apparent instances of travel-to-the-past, would it be more accurate to categorize the TWIDDER as someone/something from the past coming to the present? No to be confused with someone from the present traveling to the future, which may be a rare and dicey occurrence.

Around 400 A.D. St. Augustine wrote, "How can the past and future be when the past no longer is, and the future is not yet? As for the present, if it were always present and never moved on to become the past, it would not be time but eternity." (Well, THAT clears it all up!)

So the past is gone, the future doesn't exist, and the present exists only for an instant. But it's all one anyway, so why split hairs?

Would this be a good time to revive the sock drawer analogy? Or not?

Moving on . . .

On occasion, the experiencer brings back tangible evidence of his/her remarkable trek: a lace-edged handkerchief from the almshouse, envelopes from a stationery shop, even a casing from an old war plane.

THE CASE OF THE CURIOUS SHELL CASING

Martin Caidin (1927-1997) was an American author and authority on aeronautics and aviation. Caidin wrote more than fifty books, including *Samurai!, A Torch to the Enemy,* and many other classic works of military history.

Caidin also recorded an odd occurrence experienced on an airborne transport in 1961. He and a cadre of buddies were flying three aged Boeing B-17G Flying Fortresses from Arizona to England. The bombers were scheduled to be used in the motion picture, *The War Lover.*

Just prior to crossing the Atlantic, the crew took on one more old war colleague, Bert Perlmutter.

In his book *Ghosts of the Air,* Martin Caidin's (1994) words tell a fascinating TWIDDER tale.

We took off from Teterboro, [New Jersey] in a formation of two, leaving one B-17G behind for some mechanical work to catch up with us later. It was a beautiful flight. Because there wasn't any need for machine-gun mounts or the waist guns themselves [anymore], the two waist-gunner positions had been sealed with Plexiglas.

It was warm, and the Fort bounced and rocked gently as we droned [on]. I had come back from the cockpit to talk with Bill Mason and Jim Nau, who were photographing and recording the salient points of the flight. We sat on crates and a tire, relaxing. Looking back from this position toward the tail turret, the interior of the Fortress had a misty quality, the result of dust bouncing from aircraft motion and the sun slanting in at an angle.

Then one man gasped as he looked back through the fuselage.

'Holy...' The words failed to come. He stared wide-eyed at me and Bill Mason. 'I can't believe . . . Am I really seeing what I think I'm seeing?'

Caidin grabbed Mason's arm and instructed everybody to write down what they were seeing without discussing it. He wanted NO ONE affecting the thoughts of anyone else.

On the ground that night they compared notes. Everything written down by each man was almost a perfect match.

In his own words Caidin continued his tale.

In that ethereal light we saw dim shadows moving. Two men in heavy flying suits, wearing oxygen masks, each with his hands on a fifty-caliber machine gun in the waist position.

Calling out to one another, the guns visibly hammering, shell casings flying through the air . . . we even saw the shell cases gleaming in the shafts of sunlight.

Then one man lurched and moved with great effort, coming toward us, unrecognizable because of his leather helmet and oxygen mask. He supported another figure, who seemed to be barely conscious.

It was difficult to make out any detail, but then—and we all wrote it down—we saw that one hand of this second man had been blown away. At the wrist there was only a stump.

The other [man] was half dragging him to the open space of the navigator's upward gun position where a single machine gun was mounted to fire upward.

This was behind the power turret just aft of the cockpit. We watched the man in the mask heave the other upward and shove the blood-spurting arm into the screaming windblast. At four or five miles high, the temperature would have been forty or fifty degrees below zero—just what was needed to close off the ghastly wound and freeze the arm to stop the spurting blood.

Caidin then shared the REST of the story. Just before strapping in for landing, Bert Perlmutter moved across the fuselage where something on the floor of the Fortress caught his eye.

The three of us watched him lean down and pick up an object. He held it up, studying it in the bright shaft of sunlight. We all saw it clearly.

The casing of a fifty-caliber round. Shiny. Brand new. It wasn't in the Fortress when we took off from Teterboro. Bert turned it round and round, shook his head slowly, and placed the casing in a pocket.

THE HANDKERCHIEF

In Corrine Kenner and Craig Miller's *Strange but True* (1997) story, Elsie Hill shared an experience from August 1964. At that time she and V. Stephens, both British, holidayed in Bruges, Belgium.

One day we toured the town in a horse-drawn cab and stopped at several places of interest, among them the Old People's Almshouses.

We found ourselves in a square surrounded on three sides by cottages with a plot of untilled ground in the center. Elderly people were sitting around talking and making lace at little tables.

One lady invited me into her cottage and offered me orange squash. My friend bought a lace-edged handkerchief. We visited the chapel which was on the left-hand side of the place halfway down the block, and then we said farewell and left.

We vacationed in Bruges for a week, and before we departed, we decided to have another horse-cab ride. We asked the driver to stop at the Almshouse. He took us through the familiar gateway, and we found ourselves once again in the square.

But the scene was different from the one we remembered. There were no people about. We saw no lace makers, even though it was a warm day. The large, untilled plot of ground now held a mass of fully grown flowers and vegetables. We went to look for the chapel, but it was not there. Eventually we found it at the end of the block.

We met an attendant and told him that we thought the chapel had been on the left in the center of the left-hand row of cottages.

'Oh, yes, it used to be,' he replied. 'But it's been moved.'

We left somewhat mystified. Where were the people and the lace makers? Why was every door shut when previously each had stood hospitably open? How was it that the open field in the center had been cultivated when a week ago it had been bare? And how did the chapel get moved and rebuilt in the space of a few days?

Skeptics might say we dreamed it all, but my friend still has the handkerchief she bought that day, and that is real enough (pp. 35-36).

A "STATIONERY" MOMENT IN TIME

In *The Directory of Possibilities*, edited by Colin Wilson and John Grant (1981) is found an unusual story about a man named Mr. Squirrel and a certain shop. In 1973 Mr. Squirrel (I kid you not) went into a stationer's shop in Great Yarmouth, England to buy some envelopes.

He was served by a woman in Edwardian dress from whom he bought three dozen envelopes for a shilling. He noticed that the building was extremely silent. He heard no sounds of traffic outside.

On visiting the shop three weeks later, he found it completely changed and modernized. The assistant, an elderly lady, denied that there had been any other assistant in the shop the previous week.

Joan Forman, author of *The Mask of Time*, interviewed Squirrel and then conducted her own investigation of the occurrence. Upon contacting the stationery manufacturers, she learned that they had ceased manufacture of the particular envelopes Mr. S. had purchased fifteen years earlier!

THE SHOP THAT WASN'T

"What's in Store: Transmuting Shop Baffles U.K. Police" (n.d.) is the tale of Frank, a Merseyside, England police officer, who took a rather unexpected trip one off-duty day in July, 1996, in Liverpool's Bold Street area.

Frank and his wife, Carol, were in Liverpool one sunny Saturday afternoon shopping. At Central Station, the pair split up; Carol went to Dillon's Bookshop to buy a copy of Irvine

Welsh's *Trainspotting*, and Frank went to HMV to look for a CD he wanted.

Shortly into his stroll to the music shop, he walked up the incline near the Lyceum Post Office/Café building that leads onto Bold Street, when he suddenly noticed he had somehow entered an oasis of quietness.

Suddenly, a small box van that looked like something out of the 1950's sped across his path, honking its horn as it narrowly missed him. Frank noticed the name on the van's side: Caplan's. When he looked down, the confused policeman saw that he was standing in the road. Frank crossed the road and saw that Dillon's Book Store now had "Cripps" over its entrances. More confused, he looked in to see—not books—but women's handbags and shoes.

When he looked around, Frank realized people were dressed in fashions from the 1940's. Suddenly, he spotted a young girl in her early 20's who was dressed in a lime-colored sleeveless top. The handbag she was carrying had a popular modern-day brand name on it, which reassured the policeman that maybe he was still partly in 1996. It was a paradox, but the policeman was slightly relieved. He smiled at the girl as she walked past him and entered Cripps.

As he followed her, the whole interior of the building completely changed in a flash to that of Dillon's Bookshop of 1996.

As she was leaving the store, Frank lightly grasped the girl's arm to attract attention and said, "Did you see that?"

She replied, "Yeah! I thought it was a clothes shop. I was going to look around, but it's a bookshop."

It was later determined that Cripps and Caplan's were businesses based in Liverpool during the 1950's.

EXIT, STAGE LEFT

In *The Encyclopedia of the World's Greatest Unsolved Mysteries*, edited by John Spencer (1999), is found a story about

Vera Conway who traveled to London for a music lesson. She went to the first floor of the building to which she'd been directed. Not correctly understanding the directions, she entered a door between two cloakrooms.

Vera found herself in a theatre but realized that the audience was in period dress, possibly from the Regency period. A man wearing breeches and powdered hair approached her.

There were no electric lights, just lanterns. Vera felt strange, but no one appeared to notice anything amiss about her.

Realizing all was not right, she left the theatre and returned to the reception area to ask again for directions. She later returned to that corridor and determined for certain that there was **no** door between the two cloakroom doors!

VERSAILLES TWIDDER

The following account from Johannes von Buttlar's *Time Slip—A Parallel Reality* (1978) is one of the oldest, and most venerable recorded time-slips.

In August 1901, two Englishwomen visited Paris. They were Annie Moberly, principal of St. Hugh's College in Oxford, and a colleague, Dr. Eleanor Frances Jourdain. After a short stay in the capital, they continued to Versailles.

The two Englishwomen visited the palace at Versailles. After touring the building itself, they descended the steps into the gardens, walking toward the Petit Trianon. There they turned off along a track and passed by some deserted farm buildings in front of which there was an old plough. On the path stood two men in long green coats who were wearing three-cornered hats.

Eleanor Jourdain asked for directions, and they replied with dignified gestures. The two Englishwomen gathered that they should continue walking in their current direction. They went on their way without giving another thought to the strangers' period costume, assuming it to be intended as a tourist attraction.

They strolled up to an isolated cottage where a woman and a twelve or thirteen year old girl were standing at the doorway.

Both were wearing white kerchiefs fastened under their bodices.

Eleanor Jourdain noted that the woman was standing at the top of the steps, holding a jug and leaning slightly forward. The girl stood beneath her, looking up at her and stretching out her empty hands.

"She might have been just going to take the jug or have just given it up. I remember that both seemed to pause for an instant as in a motion picture," Dr. Jourdain would later write.

The two Oxford ladies went on their way and soon reached a pavilion that stood in the middle of an enclosure. The atmosphere was depressing and unpleasant.

A man, wearing a coat and a straw hat, was sitting outside the pavilion. His face was disfigured by smallpox. He seemed not to notice the two women; at any rate, he paid no attention to them.

Suddenly, a young man in a dark coat and buckle shoes appeared and ran past shouting something like, "You can't go through there." He pointed toward the right and added, "You'll find the house over there."

Although the Englishwomen spoke French, they could only partly understand the man's speech. He bowed with a curious smile and disappeared. The sound of his hurrying footsteps hung in the air for a long time.

The Englishwomen walked on in silence and after awhile reached a narrow, rustic bridge, which led over a ravine. A small waterfall made its way down a slope, flowing and bubbling between stones and fern leaves.

On the other side of the bridge, the path wound along the edge of a meadow surrounded by trees. In the distance stood a small country house with shuttered windows and with terraces on either side.

A lady was sitting on the lawn with her back to the house. She held a large sheet of paper or cardboard in her hand and seemed to be working at or looking at a drawing. She was no longer in the bloom of youth but still looked most attractive. She wore a summer dress with a long bodice and a very full, apparently short skirt, which was extremely unusual. She had a pale green fichu/kerchief draped around her shoulders, and a large white hat covered her fair hair.

At the end of the terraces was a second house. As the two women drew near, a door suddenly flew open and slammed shut again. A young man with the demeanor of a servant, but not wearing livery, came out.

Since the two Englishwomen thought they had trespassed on private property, they followed the man toward the Petit Trianon. Quite unexpectedly, from one moment to the next, they found themselves in the middle of a crowd—apparently a wedding party—all dressed in the fashions of 1901.

The two Englishwomen took the coach from the palace back to their hotel and started their journey home.

On their return to England, Annie Moberly and Eleanor Jourdain discussed their trip and began to wonder about their experiences at the Petit Trianon.

They decided to investigate the matter in detail.

They analyzed the events of the afternoon of 10 August 1901, at the Petit Trianon; the unusual costumes of the people they had met, and the inexplicable uneasiness that had overcome them.

In July 1904, the two Englishwomen returned to Versailles. They discovered that the cottage outside which Dr. Jourdain had seen the woman and the girl looked totally different. And the place where they had met the two men in 18th-century costume was also completely changed. The path on which the man had shown them the way was no longer to be found.

In fact, all the features of the landscape seemed to have changed. There was no wooden bridge and no waterfall, and in the place where they had seen the lady sitting in the meadow, a bush was growing. The house on the terraces, too, did not remotely resemble the one that they had seen three years before.

Faced with all these anomalies, the Englishwomen decided to undertake a systematic investigation. The task took them several years. They procured old maps and plans of Versailles and its surroundings, examined documents in the Bibliotheque Nationale in Paris and enlisted the help of historians. Gradually, a clearer picture began to emerge.

The plough that Eleanor Jourdain had seen, for example, did not belong to the Petit Trianon, but there were records to show that it had once been kept there and had been sold after the French Revolution.

In 18th-century Versailles, the only people who wore green livery were royal servants. The two men in green coats could be identified as the Bersy brothers, who had been on watch on 5 October 1789, the last day that Marie Antoinette spent at the Petit Trianon.

The cottage was shown on an old map near the entrance to the Petit Trianon. A general plan of Versailles in the year 1783 showed that a round pavilion with pillars and the Temple d'Amour had existed around the time of the French Revolution. In fact, the Temple d'Amour still existed.

Both the girl and the pockmarked man were identified from historical sources. The fourteen year old girl was the gardener's daughter, Marion, and the man with the straw hat over his pockmarked face was Count de Vaudreuil, a Creole who had played a significant part in the downfall of Marie Antoinette. In 1789, the sombrero was just coming into fashion.

According to historical sources, the running man with the buckle shoes must have been de Bretagne, a page who was sent by the palace's majordomo to the Trianon with an urgent message for the queen. He was to tell Marie Antoinette to escape immediately because the mob was already on its way to Versailles from Paris.

The door that had banged shut behind the servant had been nailed up since the French Revolution. The man was possibly LaGrange, the doorkeeper.

The Englishwomen also discovered from the historical sources that the queen had been in the gardens on 5 October 1789 when the messenger brought her the news that she should return directly to the Trianon, from where she could be brought to safety. Having delivered his message, the man ran straight off to fetch a coach.

The archives even contained the name of the dressmaker who worked for the queen. She was called Madame Eloff, and it appeared that in the year 1789 she had made two green silk fichus for Marie Antoinette.

In 1902, Annie Moberly happened to set eyes on a portrait of the queen painted by Wertmuller and was amazed to find that it had the features of the lady in the meadow near the Trianon.

It is interesting to note that on 10 August 1901, the day of their experience, electrical storms were recorded over Europe, and the atmosphere was laden with electricity. Could this have led to an alteration in the local temporal field around Versailles?

MORE MISPLACED COTTAGES

Mr. P. Chase of Surrey, England, reported that in 1968 he was waiting for a bus in the Surrey countryside. Getting bored, he went for a stroll and found two thatched cottages with a sign on the wall dating them to 1837. He admired their beautiful gardens.

Later at work Chase mentioned these to a local man, and an argument ensued because the man said there were only two modern brick houses at that spot.

Mr. Chase refused to accept this, and that evening he took the same walk. There were no cottages, just the brick houses.

Investigations established that thatched cottages had once stood on the site but had been demolished early in the twentieth century.

THE HOTEL

In her book *Time Storms*, Jenny Randles (2002) shared this unbelievable tale. It all began innocently enough in October 1979 when two couples from Dover, England, set off on vacation together. Their itinerary included travel through France and Spain. It ended in a journey that took them to another world.

Geoff and Pauline Simpson and their friends Len and Cynthia Gisby boarded a boat that took them across the English Channel to the coast of France. There they rented a car and proceeded to drive north. Around 9:30 that evening, October 3, they began to tire and looked for a place to stay. After pulling off the autoroute, they saw a motel.

Len went inside and in the lobby encountered a man dressed in an odd plum-colored uniform. The man said there was no room in the motel, but there was a small hotel south along the road. Len thanked him and he and his companions went on.

Along the way, they were struck by the oddness of the cobbled, narrow road and the buildings they passed. They also saw posters advertising a circus.

"It was a very old-fashioned circus," Pauline recalled. "That's why we took so much interest."

They drove further down the road until they saw two buildings: one, a police station, the other, an old-fashioned two-story building bearing a sign marked "Hotel." Inside everything was made of heavy wood. There were no tablecloths on the tables, nor was there any evidence of modern conveniences, such as telephones or elevators.

The rooms were no less strange. The beds had heavy sheets and no pillows. There were no locks on the doors, only wooden catches. The bathroom the couples had to share had old-fashioned plumbing.

After they ate, they returned to their rooms and fell asleep. They were awakened when sunlight filtered through the windows, which consisted only of wooden shutters—no glass. They went back to the dining room and ate a simple breakfast with "black

65

and horrible" coffee, Geoff remembered.

As they were sitting there, a woman wearing a silk evening gown and carrying a dog under her arm sat opposite them.

"It was strange," Pauline said. "It looked like she had just come in from a ball, but it was seven in the morning. I couldn't take my eyes off her."

At that point, two gendarmes entered the room.

"They were nothing like the gendarmes we saw anywhere else in France," noted Geoff. "Their uniforms seemed to be very old. The uniforms were deep blue and the officers were wearing capes over their shoulders. Their hats were large and peaked."

Despite the oddities, the couples enjoyed themselves. When they returned to their rooms, the two husbands separately took pictures of their wives standing by the shuttered windows.

On their way out, Len and Geoff talked with the gendarmes about the best way to take the autoroute to Avignon and the Spanish border. The officers didn't seem to understand the word "autoroute," and the travelers assumed they hadn't pronounced the French word properly. In fact, the directions they were given were quite poor; they took the friends to an old road some miles out of the way. They decided to use the map instead and take a more direct route along the highway.

After the car was packed, Len went to pay his bill and was astonished when the manager asked for only 19 francs. Assuming there was some misunderstanding, Len explained that there were four of them, and they had eaten a meal. The manager only nodded. Len showed the bill to the gendarmes, who smilingly indicated there was nothing amiss. He paid in cash and left before they could change their minds.

On their way back from two weeks in Spain, the two couples decided to stop at the hotel again. They had had a pleasant, interesting time there, and the prices certainly couldn't be beat. The night was rainy and cold and visibility poor, but they found the turnoff and noticed the circus signs they had seen before.

"This is definitely the right road," Pauline declared.

It was, but there was no hotel alongside it. Thinking that somehow they had missed it, they went back to the motel where the man in the plum-colored suit had given them directions. That motel was there, but there was no man in the unusual suit and the clerk denied that such an individual worked there.

The couples drove three times up and down the road looking for something that they were now beginning to realize was no longer there!

They drove north and spent the night in a hotel in Lyons. A room with modern facilities, breakfast and dinner cost them 247 francs.

The couples kept quiet about their experience for three years. They only told the story to friends and family. One friend found a book in which it was revealed that gendarmes wore the uniforms described prior to 1905.

The travelers? They have no explanation for what they experienced.

"We only know what happened," says Geoff.

FATHER, THEN SON TWIDDERS

Derek E. (2002) wrote about a time slip he experienced.

My dad was a taxi driver in Glasgow, Scotland, when I was a kid. Taxi drivers always have stories about famous people they had in the cab or crazy hires they got, or great ones they just missed—and he had the lot of them. He also had one that was a bit more unusual.

One day, in, I suppose, the late 1960's, he was driving in the north of the city along Maryhill Road near Queen's Cross, one of the older parts of town and once its own separate community outside the city.

67

One minute it was 'now'—cars, buses, modern clothes, tarmac roads, etc., and the next he was in some earlier time. It was certainly pre-Victorian given the clothes he described people wearing, horses, rough road, lower building, people in rough clothes and bonnets, etc. It lasted as long as it took him to be aware of it, and then it vanished and he was back in *now*.

The only thing I can think of that happened to me that was in any way similar was about 20 years ago (1982), when my then-wife and I were on a driving holiday in the North York Moors in England—you'd know it as American Werewolf in London country.

We went to a tiny coastal village called Staithes, which had a steep winding and narrowing road down to the harbour with the entrance to the houses and narrow footway at a higher level, say three or four feet.

We parked at the top of the village—hamlet really—where the tourist buses and cars had to stop and made our way down on foot.

What I remember is a brilliantly sunny day with lots of other people around, but as we made our way down, it just suddenly seemed as if no one else was there but my wife and me.

An old woman appeared on the footway opposite us. It became cooler and duller. She asked in what seemed to me an old-fashioned and very polite way (but may just have been the local accent and dialect) what year it was.

Now lots of old people get confused and it could have been that, but what I remember vividly is her black clothes—handmade, rough and with hand-sewn buttons—really big compared with modern ones.

Her shoes were very old fashioned with much higher, chunkier heels than you'd see an older person wearing nowadays. In the time it took me to turn to my wife and say, 'Did you see that?' she was gone. The sun was back and so were all the people. She had however, seen the same old woman and felt the same chill."

THE OLD CAR

Tim Swartz (2002) told of a person named Lyn who lived in a small Australian outback town in 1997. The town had been built in 1947 and had changed little since that time.

I was driving toward the main intersection of the town when suddenly I felt a change in the air. It wasn't the classic colder feeling but a change like a shift in atmosphere. The air felt denser somehow.

As I slowed at the intersection, I seemed to be suddenly transported back in time to approximately 1950. The road was dirt, the trees were gone and coming toward me to cross the intersection was an old black car, something like a Vanguard or old F.J. Holden.

As the car passed through the intersection, the driver was looking back at me in total astonishment before he accelerated. From what I could see, he was dressed in similar 1950's fashion, complete with hat.

This whole episode lasted perhaps 20 seconds and was repeated at least five times during my time there, always at the exact spot. I tried to make out the registration plate number, but the car was covered in dust.

RAVENNA, ITALY

Carl Gustav Jung was a Swiss psychiatrist, an influential thinker and the founder of analytical psychology or Jungian Analysis. Jung's most famous concept was the "collective unconscious." Much of the work he did was important for measuring people's personality traits. The oft-used Myers-Briggs Personality Inventory is based on his ideas.

In 1913 Jung visited the ancient town of Ravenna, Italy, and in 1933 he revisited the tomb of Galla Placidia in Ravenna. Following is an account, in his own words, of a TWIDDER.

I was there with an acquaintance, and we went directly from the tomb into the Baptistery of the Orthodox. Here what struck me first was the mild blue light that filled the room, yet I did not wonder about this at all. I did not try to account for its source, and so the wonder of this light without any visible source did not trouble me. I was somewhat amazed because in place of the windows I remembered having seen on my first visit, there were now four great mosaic frescoes of incredible beauty which it seemed I had entirely forgotten. I was vexed to find my memory so unreliable.

The mosaic on the south side represented the baptism in the Jordan. The second picture, on the north, was the passage of the Children of Israel through the Red Sea. The third, on the east, soon faded from my memory. It might have shown Naaman being cleansed of leprosy in the Jordan; there was a picture on this theme in the old Merian Bible in my library, which was much like the mosaic. The fourth mosaic, on the west side of the baptistery, was the most impressive of all. We looked at this one last. It represented Christ holding out his hand to Peter, who was sinking beneath the waves. We stopped in front of this mosaic for at least 20 minutes and discussed the original ritual of baptism, especially the curious archaic conception of it as an initiation connected with real peril of death. Such initiations were often connected with the peril of death and so served to express the archetypal idea of death and rebirth. Baptism had originally been a real submersion which at least suggested the danger of drowning. I retained the most distinct memory of the mosaic of Peter sinking, and to this day can see every detail before my eyes: the blue of the sea, the individual chips of the mosaic, the inscribed scrolls proceeding from the mouths of Peter and Christ, which I attempted to decipher.

After we left the baptistery, I went promptly to Alinari to buy photographs of the mosaics but could not find any. Time was pressing—this was only a short visit—and so I postponed the purchase until later. I thought I might order the pictures from Zurich. When I was back home, I asked an acquaintance who was going to Ravenna to obtain the pictures for me. He could not locate them, for he discovered that the mosaics I had described did not exist.

The memory of those pictures is still vivid to me. The lady who had been there with me long refused to believe that what she had seen with her own eyes had not existed. As we know, it is very difficult to determine whether, and to what extent, two persons simultaneously see the same thing. In this case, however, I was able to ascertain that at least the main features of what we both saw had been the same.

This experience in Ravenna is among the most curious events in my life. It can scarcely be explained. A certain light may possibly be cast on it by an incident in the story of Empress Galla Placidia (d. 450). During a stormy crossing from Byzantium to Ravenna in the worst of winter, she made a vow that if she came through safely, she would build a church and have the perils of the sea represented in it. She kept this vow by building the basilica of San Giovanni in Ravenna and having it adorned with mosaics. In the early Middle Ages, San Giovanni, together with its mosaics, was destroyed by fire, but in the Ambrosiana in Milan is still to be found a sketch representing Galla Placidia in a boat (pp. 284-286).

GREECE, CRETE, AND, UH, GREECE AGAIN!

British historian Arnold Toynbee developed what is perhaps the most famous of the cyclical theories of historical change in his monumental 12-volume work *A Study of History* (1934-1961). In 1912, at the age of 23, he visited Greece and had several (!) TWIDDERS.

On January 10 of that year, Toynbee was perched on one of the twin summits of Pharsalus and staring out over the sunlit landscape and thinking about the battle that took place there in 197 B.C.

Suddenly he slipped into what he termed a "time pocket" and found himself back in the days when the forces of Philip of Macedon faced the Roman legions at this spot.

The weather had changed. There was a heavy mist that parted to allow him a view of the downhill Macedonian charge that led to disaster. The Romans spotted a weakness in the Macedonian flank, so they wheeled their men and attacked with such ferocity that Toynbee had to turn his face away from the slaughter. Almost at once the scene disappeared, and Toynbee was back in a peaceful, sunlit present.

Two months later it happened again, this time on Crete. He was at the ruins of a mountain villa when with much the same sensation as an aircraft hitting an air pocket, he dropped into another time pocket.

This one jumped Toynbee back just 250 years to the day when the house had been abruptly abandoned.

Yet again (talk about winning the lottery!) Toynbee hit a time pocket in the Anatolia region of Greece at the ancient site of Ephesus. He was inspecting the ruins of the open air theatre where Demetrius approached the apostle Paul in New Testament times. Paul had made himself pretty unpopular with statuary merchants by admonishing locals to not buy the little silver models. Demetrius, a silversmith, was embittered by Paul's efforts. Finding his trade diminished through the spread of Christianity and the decline of heathen worship, he and his fellow-craftsmen instigated the uproar as recorded in Acts 19. At one point the crowd was "full of wrath, and cried out, saying, 'Great is Diana of the Ephesians'" (Acts 19:28).

Toynbee saw it all. He was even able to pick out Paul's two companions, Gaius and Aristarchus. As the chant of "Great is Diana of the Ephesians" died away, he transitioned back to the present.

UNFAMILIAR TERRITORY

Tom Slemen in *Haunted Cheshire* (2000) disclosed the TWIDDER of a successful entrepreneur and experienced pilot named Davies who took off from a private airfield on the outskirts of Chester, England, in the summer of 1992.

Mr. Davies was at the controls of a Cessna and was headed for Liverpool's Speke Airport. As the Cessna was passing over the Stamford Bridge area, Mr. Davies noticed something quite strange.

Thousands of feet below there were no signs of the M53 or M56 motorways. Nor was there any sign of a single A or B road.

Intrigued and somewhat alarmed at the apparent missing roads, Mr. Davies descended in his plane to take a closer look at the now unfamiliar territory.

What he saw made his heart somersault: a tight formation of men were marching down a road towards a long, rectangular, squat building. Mr. Davies located a pair of binoculars and trained them on the marching figures. They were Roman soldiers, and the building they were walking towards looked like some sort of Roman villa.

At this point, a strange low mist materialized and enshrouded the landscape below. When it cleared, there was the A548 motorway, and Mr. Davies saw to his relief that everything had returned to normal.

He flew over the M56 and decided to circle back to see if he could get a glimpse of the soldiers again, but they were nowhere to be seen.

Mr. Davies subsequently made extensive inquiries to ascertain if the soldiers had been extras in some film. However, there were no films about Romans being shot anywhere in Cheshire or anywhere in Britain, for that matter.

Mr. Davies realized that he had somehow ventured into the airspace of Cheshire during the Roman occupation of over a thousand years ago.

THE KIDS

On the paranormal phenomenon website, Ronnie M. (2009), shared his time slip about when he lived in London. On a Saturday in the latter part of October 1969, he was walking home late.

> I had to walk through an underpass, which was under the busy North Circular Road. It was cold and late, and I was surprised to see about five kids down there collecting pennies for the 'Guy' [since] firework night, November 5th, was soon.

Until recently, a common sight in English streets in the weeks leading up to November 5 was of youngsters standing beside home-made effigies of Guy Fawkes, demanding, "A penny for the Guy!" Bonfire night was a children's festival when boys and girls built their own bonfires and set off fireworks. The Fireworks Act of 2003 prohibited people under the age of eighteen from carrying fireworks in public. Bonfire night is now strictly controlled by grown-ups.

These kids should not have been out that late, seeing as the oldest was a girl aged about 12 years old and the others younger.

What shocked me was their clothes. Their attire made me think they had come straight out of the 1920's or 1930's London. Their speech could have been taken straight from a Charles Dickens' novel.

I heard one young boy say, 'That other gent gave me a florin.' At his age there is no way he could have known what a florin was, an old English coin for, at the time, two shillings. This was the late 1960's, and kids certainly didn't use words like 'gent' anymore. 'Geezer,' or 'bloke' perhaps.

The girl approached me saying, 'Evening sir, penny for the Guy, please, sir?' Her politeness shocked me, but I said I hadn't any money. She slid her arm through mine, and she ran her hand down my sleeve saying, 'Yes you do, sir. You are a fine gent. You do have money.' I assured her I hadn't and I expected a rude mouthful, but she replied, 'Ok, thank you, sir. You have a good evening, sir.'

I knew I had to give these kids something, so I pulled a silver sixpence from my picket and called her. I threw her the coin, and she gave me a thank you and a beaming smile. I walked off into the night.

This experience bugged me bad. Who were those kids from the past? I asked local people if any kids were killed there during World War II, but nobody remembered. Did I meet kids from the past? I guess I will never know.

MASEFIELD AND MONTROSE

In another story from Martin Caidin's ghost book, we meet Sir Peter Masefield, one of the best-known aviators in the United Kingdom. On May 27th 1963, taking advantage of unexpectedly beautiful spring weather, Masefield was flying his personal Chipmunk, a single-engine, low-winged monoplane generally accepted as one of the finest types of its class in the world.

Masefield's route from Dalcross to Shoreham brought him close to the now-abandoned airfield of Montrose. Masefield knew Montrose well. He eased into a gentle turn and was soon flying along the seashore, cruising at twenty-five hundred feet.

In the distance Masefield saw the old runways appear. He thought of the tens of thousands of takeoffs and landings of the past, and how swiftly an abandoned field becomes lifeless.

Ahead of Masefield there appeared another plane. He judged it to be at his own altitude and was surprised to see an unusually shaped biplane--not that biplanes were rare in England. In fact, many of the flying clubs banged about in the old de Havillands and other machines.

But this was no ordinary biplane. He looked with great surprise and some disbelief, for he was closing rapidly on a truly ancient machine—a seventy-horsepower BE-2 trainer that the Royal Flying Corps had placed in use before World War I.

Sir Masefield flew close enough to the other machine to see that the pilot was wearing a leather helmet and goggles and the flying scarf so popular in ancient flying days. For a few moments he wondered if the BE-2 had been a rebuild project. Flying enthusiasts were rebuilding all sorts of machines these days from past wars, but Masefield had never heard of anyone getting one of these rattletraps back into the air.

Masefield's attempts to make sense of what he was seeing were interrupted by what happened next; the outer part of the upper right wing broke within the structures. The outer wing section lifted like a loosely flapping rag.

The entire right upper wing followed in the same manner, wrenching free of its strut supports. Immediately the ancient BE-2 whirled crazily out of control as the wind hurled about the tangling mass of structure and fabric.

The aircraft staggered suddenly and fell, twisting and spinning, almost straight down to crash on the abandoned Montrose aerodrome.

Shaken by what he'd seen and helpless to offer any aid from aloft, Masefield dove earthward and set up a frantic short landing on a long golf-course green that paralleled the old airfield. He shut down his engine and dashed from the Chipmunk, shouting to several players on the green to come to his aid.

For several moments they stared at Masefield. No one had seen the plane crash, but urged on by his sharp insistence; they hurried across the field to the crash site.

There was no wreckage. There was neither a BE-2 nor the crushed remains of any airplane!

As Masefield related the specific details of what he had seen from the air and what he had failed to find on the ground, he repeated that he *had* seen the airplane collapse in flight, *had* seen it crash, and *had* seen the dust thrown up by the impact. Later he did some checking.

What Masefield witnessed appeared in the findings of the Accidents Investigation Committee of the Royal Aero Club, dated 2 and 10 June 1913. Almost fifty years to the day before Masefield's flight, on May 27, 1913, the BE-2 biplane of Lieutenant Desmond L. Arthur folded its upper right wing and plummeted to the ground.

Montrose Aerodrome remains closed.

A REUNION?

O'Neill in "Time Slip in Australia" (October 2003) narrated the story of his TWIDDER. Around 1995/96 Mr. O'Neill, his wife and two children traveled from the north of South Australia to the Barrossa Valley of Australia. This was a regular trip for the family. His wife and kids were soon asleep. "Due to the amount of extended traveling we did, I developed a driving habit of picturing the next few kilometers and checking side roads, etc.

Once one section was completed, I would concentrate on the next. This would keep me awake and alert for safe driving.

I was entering a straight level stretch of road from some hills. I could see the small town ahead called Allen-Dale North (basically a small stop with a hotel on the main road), which is just outside of Kapunda.

Looking ahead, I could see a large group of people milling about the hotel. This interested me as there had only ever been a couple of cars outside the hotel at anytime when I had traveled through.

As I got closer, I could naturally see this group in closer detail. These people were at a fair of some kind; I assumed that this gathering must have been a reunion of some sort. Every person was dressed in early 1800's style. Very English/Victorian era dress sense [clothing].

I needed to slow down and recall that on my left was a horse and buggy. As I traveled through the group, there was a boy about seven years old holding onto a woman's hand. I presumed she was his mother.

He watched intently as I approached and drove past. I watched him as he was close to the road. He was dressed in a navy blue sailor outfit with white frills. He had the most penetrating blue eyes.

No one else seemed to notice my travel through the group, but his eyes did not leave me. I thought the authorities would have closed the road off due to the number of people walking across the road, and I also considered stopping to check out the festivities.

I wish I had.

My wife stirred and asked where we were. I told her and suggested we stop . . . at which point I looked in the rear view mirror. And saw an empty main street!

TIME TRAPS

Once again, author and historian, Martin Caidin, shares an occurrence from *Ghosts of the Air*. In the early morning darkness of 4 August 1951, "A gap appeared to open between the present and the past." Those were Caidin's words when asked to explain just what happened on that day at the seaside village of Puys, a quiet coastal community close to Dieppe, France.

That summer morn two vacationing Brits heard the fury and sound of a battle that had occurred nine years earlier. The women, sisters-in-law who were especially close, had gone to Puys for a quiet rest.

At 4 A.M. their first "night" there, the women were startled awake by an uproar through the open window of their room. Not until later did they recall that Dieppe had been the location of a disastrous "practice invasion" of the French coast on 19 August 1942 with Puys as one of the actual landing beaches.

On that terrible morning in 1942, in a joint operation British and Canadian troops slammed into the beaches around Dieppe and into the German-occupied port itself. Dieppe lay in Normandy and would one day become one of the major landing areas of the full-scale invasion of Europe on 6 June 1944, but on this day Dieppe was to be a test, a probe; what historians preferred to call a *rehearsal* for the invasion to come.

As a rehearsal it was an unmitigated disaster. The Germans chewed the strike force to shreds. The attacking Canadian and British troops put ashore 6,086 combat troops; The Germans killed or wounded no less than 3,623 of those men.

The war ends, and time passes. On the morning of 4 August 1951, the two women were trapped in a time warp. They were bombarded by the sounds and recorded the times of events and details of what they heard.

At 4 A.M. we began to hear the shouts and cries of men as if they were shouting loudly to be heard above the sounds of a storm. Immediately thereafter in the distance, we heard gunfire, exploding shells, and then the shriek of dive-bombers plummeting earthward.

At 3:47 A.M. nine years earlier, the lead ships of the British and Canadian strike force began to exchange fire with German ships patrolling the French coast. On shore German soldiers shouted and called to one another as they rushed to their defensive positions.

At 4:50 A.M. the thundering sounds of battle heard by the two women cut off as if a switch was thrown. An uncanny silence followed.

At 4:50 A.M. the gunfire of the attacking force stopped.

The scheduled zero hour for the first troops to hit the beaches at Puys was 4:50 A.M. The attack was seventeen minutes behind its original scheduling. In the confusion of the delay, all gunfire that was scheduled to be lifted from the beach was lifted so that the invasion fleet would not fire on its own men.

The women strained to hear anything from the silence; it lasted seventeen minutes. Then precisely at 5:07 A.M. they were bombarded by long rolls of thunder, rising in crescendo to waves of explosions. They heard aircraft engines screaming as dive bombers plummeted from the skies. Against this background they could hear the faint cries of men, nearly overwhelmed by the battle clamor.

At 5:07 A.M. the first wave of assault boats and landing craft slammed into the beaches at Puys. Running late, the troops were desperate to get ashore and out from under withering enemy fire. The seventeen-minute delay was proving lethal, and the British threw everything they had into their protective barrage for the men. Swift destroyers raced perilously close to the shore to hurl direct fire at the German defenses, and waves of fighters and light bombers attacked buildings along the water's front believed to house heavy German forces at point-blank range.

At 5:40 A.M. the hotel room went crushingly silent as if once again a switch had been thrown.

At exactly 5:40 A.M. the British bombardment by warships ended.

At 5:50 A.M. a long roll of distant thunder drifted into the hotel room, growing steadily louder. With the war having ended only six years before, the women recognized the unmistakable sound of bombers flying in formation, many more bombers than they had heard prior to this moment. Beneath the massed engine thunder, they heard other confused sounds as if from a distant battle.

At 5:50 A.M. a fresh wave of British bombers with a heavy fighter escort came to Dieppe at high speed from their home bases to relieve the fighters on station. By now the Luftwaffe was out in full cry, and violent air battles spread swiftly over the beaches.

At 6:00 A.M. the sounds faded slowly from the hotel room. By 6:25 A.M. the ripping thunder of battle and aircraft engines was barely a background sound, interrupted only by barely audible cries of men. At 6:55 A.M. silence in the room.

At Dieppe and the beach landing zones, the German defenders took full control of the battle. The British and Canadians, bloodied and savaged, withdrew as fast as possible from the devastating German firepower. All survivors of the attacking force surrendered to the German army.

Postscript.

When silence fell at 6:55 A.M., the time curtain closed again. Puys and Dieppe faded forever--at least for these two incredulous women. Although the sounds of battle, cries of men, and thunder of aircraft engines at times were so loud and fearsome that they seemed to shake the entire hotel, only these two women heard anything!

D-DAY

Jamafan's (2003) TWIDDER is next. In 2001, Jamafan (he wished to remain anonymous, so he chose this name) was twelve years old. He and his folks took a trip to Normandy, France.

While I was there, I was walking down the beach by myself, the beach that was code named Sword during World War II.

As I was walking, I heard what I thought was machine gun fire in the distance. I looked around and saw nothing.

I didn't think anything of it until a few minutes later when I looked out toward the water and saw almost ghostly shapes in the water.

As they got a bit closer, I could tell that they looked like landing craft. I could see everything in detail.

I turned around to see if anyone could confirm what I was seeing, but there was no one around. I looked back out to the water, and they were much closer.

This is when I began to feel very uncomfortable, so I started to back up a little bit. I turned around for a second and then looked back and they were gone.

I have no doubt that I saw what I saw.

WORLD WAR II AND MONTROSE

In the winter of 1940, a British Hurricane fighter pilot had a most unexpected encounter. Assigned to the airfield at Montrose, Scotland, he was one of a number of squadron pilots awakened from sleep by the enemy.

The German Luftwaffe delighted in disrupting entire U.K.-based squadrons and airfields on every occasion they could manage.

That is exactly how the wintry evening began. Montrose had been selected for night harassment, and a twin-engined Heinkel bomber crossed over the field, engines thundering, disappearing into the darkness and then roaring back from what promised to be a long-delayed bomb run.

Alarms sounded in the sleeping quarters. Fighter pilots slipped into their flying gear and rushed to operations. Several Hurricanes were prepared for flight by the ground crews and held on the ground for immediate takeoff. A single pilot, an experienced flight sergeant, got the nod to get into the air immediately to find and shoot down the Heinkel. The eight-gun fighter was soon racing down the darkened runway; it lifted from the ground and disappeared into the murk, visible only by the glowing exhausts of an engine under full power.

For the next thirty minutes the seasoned sergeant-pilot did his best to find the Heinkel. For whatever reason, either low fuel or perhaps having caught sight of the Hurricane taking off, the Heinkel was gone. The radio signal went out, "Return to base and land."

In those days, no one ever knew when the Germans would play a double game. They'd send over one bomber to entice British fighters into the night sky. By the time the fighter climbed to altitude, the bomber would be gone. Landing a Hurricane at night on a dark field wasn't the simplest of tasks. However, if the British turned on runway lights, the field became a beacon for a second bomber just waiting for the chance to attack.

Stung more than once by this very tactic, the British turned on two rows of what they called glim-lights. They were so dim that only a pilot knowing where to look could find them in darkness. This pilot knew what to do, and every man on the ground could tell when the fighter's gear came down and the Hurricane was turned properly to land on the grass strip between the glim-lights.

Moments later a man called out, "There he is!" They didn't yet see an airplane, but the throttled-back exhausts cast their unmistakable glow in the night. The Hurricane came in smoothly, beginning its landing flare as the pilot skimmed over the trees at the airfield perimeter.

As they expected, the engine sounds died away to a rumble as the Hurricane was brought back to idle, and the fighter settled smoothly to the grassy surface.

Unexpectedly, with a thundering roar and a flash of bright exhausts, the pilot rammed his throttle forward to full power and raced into the air again.

For the moment, the operations staff was convinced the sergeant-pilot had caught sight of a German bomber or perhaps had seen some obstruction on the runway. Neither bomber nor obstruction could be seen. They heard the Hurricane circling the field and again setting up his approach.

This time the pilot, convinced no German bomber was about, but still catching everyone on the ground by surprise, turned on his navigation lights. Had there been an enemy aircraft in the area, the Hurricane would have made a splendid target. Something else was happening, but no one watching could figure out what it was.

The Hurricane again began its flare to land. This time the pilot went to full power before his wheels touched, and again the fighter thundered away into darkness. They caught a glimpse of the navigation lights showing the direction of flight toward and then over open sea.

Operations signaled the ground crews. Forget whatever bomber might be upstairs. "Give him a way in," they ordered. Ground crews switched on the Chance beacon, a powerful searchlight mounted in a horizontal position, and set up to illuminate the entire grass airstrip's runway in a huge glow of yellow white light.

The Hurricane came about again, crossed the perimeter, and this time the fighter touched down smoothly with power full back. Immediately the Chance light was shut down to darken the field. The Hurricane pilot taxied back using only the glim-lights, swung the fighter about, and killed his engine.

Pilots and ground crewmen crowded around the Hurricane. The sergeant-pilot slid back his canopy and stood high in the cockpit. With an angry gesture he yanked off his leather flight helmet. His voice boomed across the field.

"The fool!" he shouted, his face contorted with anger. "Who's the b----- fool who cut me out!" He climbed down and dropped from the wing.

"B----- stupid b------! He could have wrapped up the lot of us with his stupidity!"

The other pilots looked at one another and shrugged. One stepped forward. "Look, Sergeant, no one cut you out. We were watching you the whole time. You were the only machine working the airstrip."

"That's a b----- pile,' came the heated retort. "What's with you blokes? Have you all gone blind? Of course, someone cut me out!" He renewed his shouting, stabbing his hand at the runway. "Why do you think I went round again? Twice???"

An officer moved closer. "Sergeant-Pilot, what type of machine?"

"Why, sir, it was some b----- madman in a biplane. Just as I was crossing the boundary and easing into the flare, that biplane balked me. Cut before me just as I was touching down. Looked like a Tiger Moth, it did."

Silence met his words. By now the flight commander stood

before the thoroughly agitated pilot. "There's no one else flying," said the flight commander. "Besides, we don't have any biplanes on this station."

The sergeant-pilot set a famously stubborn jaw. "Sir, I know what I saw, and I saw it not once, but twice like I said, and that biplane *was right in front of me!*"

The group fell silent. Finally the flight commander gestured to his men. "That's enough for tonight. Pack it in, gentlemen."

The pilots and aircrews returned to their sleeping quarters. No one save the sergeant-pilot flying the Hurricane—and he was in the best position to see any other aircraft—had seen the mystery biplane.

THE HOUSE

May 2009, Amanda shared her experience with a shift in dimensions.

> I live in an area in Yorkshire, England, which is rural but still close to urban areas. Sometimes my husband and I go to the farm shop for our veggies, salad and meat. In order to get to the farm shop, we have to go through my sister's village and past her house, which is set next to a church.

> During the late summer of 2007, we decided to go and get some shopping. The air quality that day did seem rather different than usual in that it seemed quite thick—thick enough for us to comment that people with asthma might find it difficult to breathe.

> It wasn't particularly hot, just 'atmospheric' (for wont of a better description). It is instinct[ive] for us to look at my sister's house as we go past to see if we can see her or my brother-in-law in the garden.

On the way back from the shop, we had a very strange experience. We traveled along in silence. We didn't even have the radio on, which is unusual for us, but we were happy just thinking about nothing. I was looking out of the window to my left, ready to wave if I saw my sister or brother-in-law.

Suddenly, I felt strange inside, a weird feeling difficult to explain. A sort of whooshing sensation speeding through my body.

At the same time as this strange feeling, I noticed with panic that the road we were traveling on had slightly changed. It looked very similar, but the worst thing was that my sister's house was completely missing and the terraced houses that usually surrounded it had taken on a small but significantly different appearance!

Now all of this happened in a flash as we were still traveling in the car at normal speed, only it felt as though we were going at a hundred miles an hour for that split second. As we got past where the house should be, I screamed out in panic for my husband to stop the car. At the exact same time, my husband shouted at me, 'Mandy, where has Chris and Alex's house gone? What has just happened?'

He slammed the brakes on and pulled over onto the side. As we looked back, we could see the road had returned to normal and the house was back in place. Gobsmacked, we just stared at each other thinking, 'What the heck had just taken place?' Individually without realizing it, we had just had the exact same experience at the exact same time.

It felt to us both as if we had just taken a sneaky look at another dimension. The time period had felt the same to us and the road was the same, it had just taken on a slightly different appearance. We felt excited to have had this experience as we are both quite skeptical about such things. It was extraordinary and remains unforgotten.

Now the story deepens a bit. We did not reveal to my sister what had happened until the next day. She was shocked and then went on to inform me that on the evening of our event they had invited some friends (a couple) to their house for a meal.

The couple had been to their house many times. When the couple arrived for the meal, they told my sister and husband that they had problems finding the house. Her husband had pulled up outside where the house should have been and was told by his wife that 'Alex and Chris don't live here.'

'Don't be daft,' said her husband. 'The house should be here. What has happened?' So they drove off in a very confused state back up the road. When they came back down again, the house had reappeared!

All of this [happened] on the same day. The couple nor my sister and husband knew nothing of our story at this time. To further add mystery, my sister had seen us go past in the car that day, and she had seen the couple pull up outside and drive off again!

THE PHONE CALL

Liverpool writer Tom Slemen (n.d.), investigator of the unknown, shares a 1970's magazine article concerning a TWIDDER. Alma Bristow of Bidston, Birkenhead, England, tried to phone her sister—a recent widow—in Frodsham,

Cheshire.

Alma always had difficulty dialing numbers on the old British Telecom analogue telephone because she had stabbing arthritis in her fingers. Alma evidently misdialed her sister's number, as a man's voice answered.

The man said, "Captain Hamilton."

Alma asked if her sister was there, but Captain Hamilton replied, "This is NOT a civilian number. Who are you?"

Alma gave her name, and as she did so, she heard a sound at the other end of the phone that she hadn't heard since she was a young woman; an air raid siren.

"Sounds like World War II there," Alma joked.

There was a pause, and then Hamilton replied, "What are you talking about?"

"The air-raid siren. Sounds like the war's still on," Alma said, about to hang up.

"Of course, the war's still on. Where did you get my number?" said Hamilton.

"The war ended years ago, in 1945," said Alma, now suspecting that she was a victim of the "Candid Camera Show."

Alma could hear the captain whispering to an associate. Then he said, "It isn't 1945 yet. If we trace you, you'll be thrown into prison for this lark, you know? You're wasting valuable time, woman!"

"Eh? It's 1974. The war's been over for years," Alma retorted. Then she heard the unmistakable rumble of bombing coming over the phone.

"We'll deal with you later. Don't worry," said Hamilton, and he slammed the phone down.

Alma listened eagerly for him to pick up the handset of his telephone, but Hamilton never did. Alma never knew if she had been the victim of an elaborate hoax or whether she had really talked with someone in wartime Britain.

THE COTTAGE (Yup, yet another)

John was a six year old in Stoke-on-Trent, England, when his brush with the past occurred. He was on his way to school with his friend when they stopped to watch some builders working on some new houses.

As they approached the site, they noticed an old cottage nearby. An old lady came out and offered John and his friend some lemonade, and they went into her house.

After leaving the cottage, they continued on to school, only to discover that it was almost 4 P.M., and school was just closing. They had left home at 1:30 P.M. on a journey that should have taken about 20 minutes!

The next day, John and his friend took the same route to school, but to their amazement there was no sign of any cottage or the old lady they had seen the previous day.

The only explanation seems to be that the boys experienced some sort of time slip in which for a few hours they were transported back to an earlier time when, indeed, a cottage—and its occupant—did exist on that land.

A QUICK VISIT TO THE BRONZE AGE

L. Fanthorpe in *The World's Greatest Unsolved Mysteries* (1997) shares the tale of Mrs. Anne May, a school teacher from Norwich, who was on holiday in Inverness with her husband. They were studying the Bronze Age Clava Cairns—a small group of three burial mounds. At the end of their tour, Mrs. May rested briefly on one of the stones where she apparently experienced a time-slip.

She saw a group of men with long dark hair who were wearing rough tunics and cross-gartered trousers. They were dragging one of the massive stones.

At that moment, a party of tourists entered the site, and everything returned to normal.

HADDON HALL

Another incident in *The World's Greatest Mysteries* (1997) was reported by L. Fanthorpe. Joan Forman had a trip to the past while visiting Haddon Hall in Derbyshire.

She saw four children playing happily on the stone steps of the Hall's big courtyard. The oldest was a girl of nine or ten with shoulder length fair hair. She was wearing a greenish-grey silk dress with an attractive lace collar and a white Dutch-styled hat.

Forman could see only her back view at first, but before the time-slip experience ended, the girl had turned so that her face was clearly visible. She had broad distinctive features, a wide jaw, and an upturned nose.

As Forman moved towards the group, they vanished; it was almost as if by moving slightly, she had inadvertently turned off whatever strange current had been making the children visible.

Inside the Hall, Forman searched everywhere for a portrait of the children she had seen—especially a picture of the eldest, the girl with the unmistakably strong and distinctive face.

She found one! The greenish-grey dress was the same, so was the hat and the lace collar. The girl was identified as Lady Grace Manners, who had been connected with Haddon Hall centuries ago.

THE AMBUSH

J. H. Brennan in *Time Travel: A New Perspective* (2000) told the story of a 1940's mountaineer. In 1940 mountaineer Frank L. Smythe entered a grassy hollow near Glen Glomach in the Scottish hills. There he saw a small group of weary men, women, and children stagger into an ambush laid by men wielding spears, axes, and clubs.

In the resultant massacre, every last member of the group was slaughtered. Smythe was so horrified that he ran from the scene.

Smythe later tried to confirm through historical records what he had seen. He discovered two massacres that had happened at that spot, but the clothing and weaponry did not match what he saw in his TWIDDER.

Nevertheless, he never recanted his claim.

Unlike Smythe, the following two TWIDDER experiencers (also recorded in *Time Travel*) were able to document their historic sightings.

THE PUB

In the late 70's, Stella Knutt was on a driving tour of the Wicklow Mountains in Ireland. She turned off onto a narrow road and drove past a solitary public house, noting, among other things, a 1930's style bicycle propped against the pub wall. She then continued driving until the road changed to dirt, grass, and gravel and then seemed to disappear in the distance.

Since it was obvious she would soon be unable to drive any further, she found a turning space and came back down again. As she reached the junction with the main road, she realized she had not passed the pub.

Although she suspected she had simply not been paying attention, she was sufficiently intrigued to turn and drive back up the narrow road, this time specifically searching for the pub.

She did not find it, but she did find the open space where it had been. Half-buried in the undergrowth was the remains of a foundation which might have been the pub she saw!

THE COTTAGES

P. J. Chase, of Wallington in Surrey, was waiting for a bus one afternoon in 1968. He decided to stroll a little way down the road to pass the time.

This he managed to do in more ways than one! He came across two picturesque thatched cottages with hollyhocks in their gardens. One of them was dated "1837."

The next day Chase mentioned the cottages to a friend—only to be told they did not exist. He went to check and found the friend was right. The only buildings at the spot were two brick houses.

When he made inquiries in the area, an elderly resident confirmed that the cottages had existed. Some years previously they had been pulled down to make room for the current houses.

Chapter 4.

Forward to the Future

When last we were pondering trips-to-the-past versus trips-to-the-future, I was cross-eyed and looking for chocolate.

Time travel is rife with pitfalls.

Like paradoxes. Take the grandfather paradox. What would happen if a time traveler went back and killed his paternal grandfather before the traveler was born? How could that person then be alive to go back and slay his grandfather?

Or how about conundrums? Parallel universe theory posits alternative histories. Let's say that you travel back to meet your grandfather. In the theory of parallel universes, you may travel to another universe, maybe one that is similar to ours but has a different succession of events. Just because you kill your grandfather in another universe doesn't mean he isn't just fine in your own!

BUT, goes this line of reasoning, every time you travel to the past and diddle with history, another universe is created. One that was identical to your own up until the time you changed the original succession of events.

So are there physical laws that prevent all but the most mundane of interactions with the past? Purchasing stationery from a long-defunct shop or conversing with a gardener-from-the-past may be such trivial daily activities that there is no cause for concern? Is it that TWIDDER experiencers are in the past or future for such a transient time that mayhem is the least of their opportunities?

The pitfalls mentioned above presuppose a controlled trip to the past with the experiencer planning *when* he/she will arrive, *where* he/she will land, and *who* he/she will see. Maybe the fact that TWIDDERS are random acts outside of our control ensures us a stable present.

And why the heck would anyone want to kill his/her grandpa anyway?

FOUR YEARS IN THE FUTURE

Martin Caidin in *Ghosts of the Air* (1994) introduces us to Victor Goddard, a Royal Air Force aviator who experienced a TWIDDER. The record of Goddard's trip-to-the-future constitutes an official registry with the RAF.

It was 1934, and Goddard was caught in a heavy thunderstorm over Scotland. The British pilot flew a Hawker Hart biplane fighter, and he was in trouble.

The pilot knew that getting caught in the belly of a giant thunderstorm was asking to have his plane torn apart. He cut back on power and eased the Hawker into a glide, descending as carefully as he could, dodging and twisting through the blackest of clouds as he gave up altitude.

He knew generally where he was even though this was decades before navigation aids, such as GPS systems. Somewhere in this area was an abandoned airport known as Drem. If he could spot that field, he could whip down in a hurry and get safely onto the ground. The problem was he had to fly more by intuition than by sighting known landmarks.

Intuition paid off. He eased beneath a boiling cloud base, squeezed between the still-lowering cloud deck and increasing turbulence, and sighted the earth below. Then he recognized landmarks; if he was right, Drem would be dead ahead of his position. He brought in power and barreled for the field that promised safety. And there it was, right in front of him. He'd made it!

He was still a quarter-mile distant from the abandoned airfield and closing rapidly on his goal when the thunderstorm seemed to split wide open. The booming clouds separated, then opened wide in the form of a giant chasm in the sky. Brilliant sunlight burst through and cast a dazzling golden glow on the countryside below.

Goddard brought the Hawker over in a tight bank so that he could fly directly over Drem and check the runway for his landing. The abandoned airfield *wasn't there*. He couldn't believe what he saw; he was over Drem, all right, but the ruined field had vanished.

He was circling an airport bustling with activity. He kept his turn tight, staying within the boundaries of the beautifully laid out and maintained airfield. At a height of only fifty feet, he raced across the flight line, hangars, rows of airplanes, living quarters, vehicles, and hundreds of people—and no one even looked up at this fighter streaking by so low.

Goddard stared at the mechanics in blue coveralls working on rows of planes painted brightly yellow. That threw him for a loss. This was a training field, obviously, but the RAF trainers were all painted an unmistakable silver. And he didn't see a single silver-colored aircraft.

As he made another pass, he continued to be completely ignored from below. Goddard pushed aside his puzzlement. The clouds lifted suddenly and strangely. He recognized the landmarks about Drem airfield, and he climbed back to altitude to take up a heading that would return him to his original destination.

Now for the kicker. In 1934 Drem airfield *was an abandoned relic*. Its buildings were partially collapsed, and the runways were broken and dangerous. Not a soul lived or worked there. It had been empty for years.

In 1938, with war with Germany fast becoming a reality, the RAF returned to Drem with a crash program to rebuild the airfield and transform it into a top-priority, top-quality training installation. Soon Drem was a major RAF training base—and when it opened for full operation, *the color scheme of all the RAF trainers was changed from silver to yellow.*

Somehow, fighting his way through the stormy skies of 1934, RAF pilot Victor Goddard had slipped into a break in time/space. He had descended through a storm that mysteriously split open with brilliant sunshine flown over an airfield that would not exist for another four years!

EARLY EDITION

Ronnie M. (2009) shared a TWIDDER that resembled the television series entitled "Early Edition." The program ran from 1996-2000 and dealt with the life of Chicagoan Gary Hobson (Kyle Chandler). Somehow or another, Hobson mysteriously received newspapers (specifically, the *Chicago Sun-Times*) a day ahead of time, effectively giving him knowledge of the potential future.

His newspaper was delivered by an enigmatic tabby cat. Hobson would then try to prevent tragedies described in "tomorrow's" *Sun-Times*, which could then change the story text and headlines in the next day's newspaper (and everyone's lives) to reflect the outcome of his actions. Nuts, huh?

Well, in the mid '90's, Londoner Ronnie M. was in a local shop. He and the owner read in that day's newspaper the details of the plot of a popular soap opera that was due to air that very evening.

Except when he watched the show, it was nothing like the paper had noted. When he checked around, he could find no one else who'd read what he had.

Ronnie reconnoitered with the shop owner the next day; they took another look at the previous day's paper. Now there was no mention of what they'd both read!

Six months later, the episode the gentlemen had read about together actually aired.

If only they'd checked on the winning lottery numbers!

OZ

In October 2008, a 47 year old TWIDDER contributor (who chooses to remain anonymous), related what she experienced as a curious seven year old. She was playing alone near some woods in her neighborhood in Fitchburg, Massachusetts.

> I decided to take a walk in the woods. It was a ten minute walk, and I was walking through thick bushes or trees, pulling leaves and branches away from my body and face. I noticed the leaves were turning color from green to yellow and orange while I was walking.

> When I finally came out of the woods, I just stood at the edge. I looked ahead and stared in awe. I felt like Dorothy from *The Wizard of Oz* coming out of her tornado-dropped house and staring at the unknown.

> For what I saw was a futuristic type of neighborhood that I had never seen before. I was a little afraid because I did not know where I was.

> It was beautiful! The roads were extremely shiny and sparkling clean. I could see the sun shining off of it, modern and new looking. The roads were not made out of black tar but were silver and shiny. It could possibly have been metal of some sort.

> The houses were colorful, modern, well-kept with huge windows and different colored roofs, mainly orange. The street lights were different—straight, clear poles with an upside-down clear bowl on top of the light. I could hear humming in the background.

The cars were small and rounded; their colors were fluorescent green with black trim around the car doors and roof.

I did not see any people. I was only there for five to six minutes, just staring. Then I went back into the woods to go home. I told my mom about the beautiful neighborhood I'd seen, but she wouldn't listen to me and told me to 'stay out of the woods!'

The next day I went back into the woods to the same location, and the place was not there. I looked and looked but could not find that future place.

TRAIN? NO TRAIN?

In 2001 Mike (2006), a college senior, had a weird experience. Late one day he was returning to the dorms using the local subway.

I remember being in a hurry to get back because I was late for a dinner date. As I waited, I looked anxiously down the tracks anticipating the train's arrival.

Suddenly I spotted the train's lights approaching. I heard the loud rumble it made, and eventually I saw it round the corner.

The next thing I recall was a waking-up type of feeling. There was no train approaching although I felt a slight breeze as if the train had just passed.

For some reason my instant reaction was to assume I somehow was daydreaming and either didn't really see a train, or it had passed without stopping, and I somehow failed to notice.

As odd as these excuses seem now, at the time I was more than willing to use these rationalizations as reasons for what had just happened.

A moment later, another train arrived, and I stepped on board. I would have totally forgotten about the event and gone about my business if it were not for some ladies talking.

One of the ladies was saying enthusiastically to the other, 'I swear it disappeared.' This perked my ears and I wandered over. As the arguing between the two continued, I chimed in.

'Excuse me,' I said, 'This may sound silly, but did you just see the train disappear?'

'YES!' she exclaimed. 'I told you so,' she told her companion.

The remainder of the ride back consisted of me and her talking about the events as we witnessed them. More people also came forward, admitting they had witnessed the same thing. What's odd is that as we talked quite loud about the strange events, other people on the train weren't even noticing what we were discussing. It was as if they were extras in a movie. I'll never forget this event as long as I live.

UP THE GARDEN PATH

On August 2007, a British gentleman wrote about an odd trip to Wales.

It was summer 2003, and we were holidaying at a caravan site in Wales. We had been down to the village to get shopping in for the week, and we were walking down the isolated lane carrying heavy bags, stopping every now and then to take photos.

Just before we took a final photo near some castle walls (the castle was now apparently converted into a hotel), we noticed a frog. Then we went under the castleway arch and carried on.

There was a dried-up pond close by, and we commented that the frog was a long, long way from water. We seemed to be walking for ages and it was hot. The path was long and uphill, much further and more hilly than either of us remembered.

Finally we saw ahead of us a woman walking with a sheepdog. She turned onto a farm road (the first turning, incidentally, that we had come across), and we caught up with her to ask the way to our caravan site.

She was astounded that we could have missed it and said we would have walked right past it and to turn back. (Just a note here, as the woman was giving us directions, a jogger ran past in the lane and glanced our way.)

Anyway, we walked back—and came to the caravan site, just by the archway as we'd originally remembered. It was complete with its usual ornamental pond, filled with water! There had been no caravan site when we walked past earlier, nothing but a dried-up pond. When we did the same walk a couple of times afterwards, the lane wasn't as long or hilly.

We often wonder if we maybe somehow went forward or backward in time. And what of the jogger? He would have been running right behind us, so did he see anything? I guess we'll never know what happened that day.

SAVED BY THE FUTURE

Californians Scott, his younger brother, and his dad were on their way to work one cold, foggy winter morning in 1993.

Driving along, in the distance they saw someone crawling along the side of the road. He had a bushy moustache and was wearing a red jacket and worn-out jeans.

They pulled over and got out to help the guy.

From the expression on his face, they could tell that he was in a lot of pain. They talked to him and found out that both of his legs were broken.

He said his car was up the road. He'd been in an accident and had been crawling for over an hour looking for help. He said he had pulled over to do some maintenance on his car when out of the fog came this big truck. It smashed the front door against his legs as he was getting out. The truck kept going; it didn't stop. It just disappeared back into the fog.

Scott's father put a blanket over the man and told him to stay there, and he would go to get help. He didn't want to take any chances of injuring the man anymore than he already was, so Scott and his dad returned to their car and headed up the road. When they had driven a short distance, they came upon the man's car. They couldn't see any damage at all on the driver's door—no broken glass scattered anywhere. It was in good condition.

Puzzled, they then noticed there was someone in the car on the driver's side. When they looked inside the car, they couldn't believe it. It was the same man! The same bushy moustache, red jacket, hair style, eyes . . . what was going on here?!

The man was getting ready to get out of the car when Scott and his father saw headlights coming toward him. Scott's father blew his horn in a long, loud blast. It distracted the man enough to make him stay in the vehicle for a few seconds. That's when the four-by-four truck came racing out of the fog, barely missing the man's car. After the truck fled the scene, they went over to see if the man was all right.

The fellow thanked them profusely; if they hadn't sounded the horn when they did, he would have stepped out of the car and gotten killed or injured. He was very grateful and drove away.

In disbelief, the good Samaritans headed back to where the injured man had been. He was gone! He hadn't crawled away because he would have left a trail or something on the dirt, but there was none. The only thing left was the blanket.

PREVIEW

A fellow using the address "xylophobia" posted his TWIDDER on the Paranormal Phenomenon website in June 2008.

> In the summer of 2001, I was shopping at the local Wal-Mart. I was wandering around the video department, killing time, while my two sons played around with the video games.

> While looking at DVD's I happened upon a movie that I hadn't thought about in a long time. The movie was *Rollerball*. I recalled watching it as a kid and was excited to see it again.

> As I picked it up, I was surprised to see that the individual actors appearing on the cover were not the ones I recalled in the movie. I didn't recall an African American in the movie as one of the key characters at all.

> On closer examination I realized that it was LL Cool J pictured on the cover. This made me immediately realize that this was not the 1975 production I'd seen as a kid.

My next thought was one of surprise that they had remade the movie, and I hadn't heard about it at all. I'm not huge on movies, but I tend to pay attention when they remake a classic I enjoyed as a kid.

With the copy of the new version in hand, I wandered around the video department and was pleasantly surprised to find that the original version with James Caan was also available on DVD.

Deciding I didn't really want to buy either movie (dumb move in hindsight), I put them back and never thought about them again after we left the store.

Never thought about them that is until January of 2002 when I was sitting at home working on the computer. TV [was] playing in the background, and a commercial came on that about knocked me out of my chair. The commercial was announcing the new movie that was about to hit theatres . . . *Rollerball*, starring, among others, LL Cool J.

Now unless Wal-Mart is heavy into video pirating (remember, I stumbled on this movie quite a long time before its release), I have never come up with a logical explanation for this event.

FUTURISTIC CITY

Daisy (n.d.) told her TWIDDER story online. She and her friend Rick were driving to another friend's house in September of 2004 in Rick's beat up old truck.

Suddenly the truck's engine died, and Rick and I were stranded on a deserted highway in the middle of the night. We were surrounded on both sides of the road by cornfields that stretched into the distance.

Rick began a desperate effort to restart the truck, but nothing seemed to work. We decided to walk to the nearest town about two miles away to find a payphone to call our friend for help. We walked for what seemed like hours, and the town was nowhere in sight.

However, just when desperation was about to grip us, we saw a light, a gloriously bright light, shining over the steep hill ahead of us. We ran up the hill that blocked us from the light, and were flabbergasted by what we saw.

Just over the hill, Rick and I saw what could only be described as a futuristic city with lights streaming out of every window of the massive, metallic towers. In the middle of the futuristic city was a huge silver dome.

I stared at the city, stunned, until Rick elbowed me, which pulled me out of my trance. He pointed to the sky. Hovering above the city were hundreds of hovercraft. One flew toward us with amazing speed.

Rick and I were so scared that we took off running back to the broken down truck. I never looked back, but I felt someone watching me the whole way.

When we got back to the truck, it started without difficulty. Rick and I took off as fast as we could in the opposite direction. We never went back or spoke of it again to this day.

HALE BANK TO BE

Tom Slemen's (2000) report of a TWIDDER also involved a brief view of the future. Between 1995 and 1997, a number of people glimpsed the local future in the vicinity of the Runcorn Bridge. (The Runcorn Railway Bridge crosses the River Mersey at Runcorn Gap from Runcorn to Widnes in Cheshire, England.) They reported seeing a breathtaking futuristic vista on the horizon towards Hale Bank near Speke Airport.

The incredible sight that greeted Frank Jones at 4 A.M. on December 5th, 1995, was not the lights of Liverpool Airport. Mr. Jones thought it looked more like Liverpool Spaceport! The scene was like something from "Close Encounters of the Third Kind." Enormous lenticular ships dotted with bright blue and red lights were rising into the dark sky until they were out of sight.

The domed ships were taking off silently from an enormous circular area which was lit with a glow from powerful lights. Dotted about this launch area, Mr. Jones could see towers and buildings speckled with myriad colored lights. As he traveled north along Queensway on the other side of the Mersey, he lost sight of the breathtaking spectacle but knew that dozens of other early morning commuters must have witnessed the same incredible scene.

FUTURE SOCIETY

Ben (2005) shared an incredible glimpse into the future in a TWIDDER may not have been just a generic trip to the future but possibly a very personal peek into his future.

One night Ben was walking with a friend on a dirt trail between the closely connected towns of Hurley, Wisconsin, and Ironwood, Michigan. And so his story begins.

We were half way to his house when suddenly I was standing outside a huge skyscraper building in what I, for some reason, believe to be Detroit.

I entered the building and there was a lady with platinum blonde hair. Her clothes didn't look odd or anything. She told me that I was on time for my appointment, so I followed her to an elevator. We stepped inside and she pushed the button for the fifty-third floor.

When we got out of the elevator, I followed her to an office. The walls and the floor were done in a decorative business-like way. We got to the door and she told me to go in and sit down.

When I went into the office, it looked huge. I don't think I have ever seen a view so panoramic and beautiful as that one.

A man told me to sit. He then started to tell me that they—they meaning the company or something—were happy that I had joined and I would be a perfect fit.

All of a sudden, I was in a good-sized hallway with about fifty other people, standing in a military-type line. We all had the same blue and black uniforms on and were marching toward a big open garage-style door.

Suddenly it all ended, and Ben was back in Ironwood, kneeling on the ground by some bushes. His friend asked him what was wrong.

'I asked how long had I been kneeling, and he said just a couple of seconds.'

Remember those prehistoric critters in Chapter 2? There was some question as to whether the folks who viewed them had trekked to the past or whether the creatures had come to the future. The following two accounts appear to be TWIDDERS of the latter.

THE MANGROVE RAPTOR

It took a father (who still wishes to remain anonymous) three years before he would share his TWIDDER on the Paranormal Phenomenon website (January 2005). You'll understand his hesitancy as you read about his memorable Christmas holiday fishing trip in 2002.

Four people and I were in an 18-foot, center-console fishing boat close by our house in a mangrove jungle on the edge of Charlotte Harbor in Port Charlotte, Florida. Our fishing party involved five guys: my son (a PFC in the 82nd Airborne of the U.S. Army at the time) and two other PFC's he'd brought home for the holidays who were stationed with him, another local man, and me.

It was approximately 2 P.M. when my son decided to have us drop him off at a nearby bridge where he could disembark and be picked up by his mom in order to visit with his fiancé, as she had just gotten off of work.

While slowly making our way in the manatee zone up a canal toward our destination bridge, we were all surprised by a very loud, raucous cry, and the strange sound of huge wings beating. Powerful whooshing sounds with a definite hard beat at the end of the down stroke echoed through the sky. I, being an amateur zoologist type, know that there is no raptor (endemic to this range or otherwise) that is not familiar to me.

The bird we saw was definitely a raptor and unknown to science. Its wingspan was approximately 12-16 feet and the bird stood about 40 inches high. The bird flew some distance, much of the time within our sight, and lit on a utility pole about half a mile away.

Its body size with wings completely folded was comparable to the standard step-down transformer it landed next to. Its color was a dull grey with a darker head and multicolored, large, hooked beak of orange and blue. There did not seem to be any featherless areas on its head or adjacent beak area.

We had a quality digital camera with us, but by the time we got it out of the waterproof storage box and ready to take pictures, the bird was at the utility pole and would have been a meaningless image showing only a tiny black dot. No one in our party (especially my son) wanted any kind of report made due to the possible tarnishing of their reputation and/or possible negative impact on future security clearances, etc.

Now, however, he's decided it's been long enough and that as long as he remains anonymous, it would be okay to report this sighting to the crypto-zoological world. Please note that there was no alcohol consumed by the witnesses over a several day period during which the creature was seen, and all the soldiers involved are of high moral and personal character.

THE BIRD THAT WAS

On a paranormal blog site, Don D. (2008) wrote about a similar sighting.

Several years ago, I believe in 2005-2006, I saw a huge bird just west of the northern side of Dade County in south Florida. I was as far back into the Everglades as you could drive.

It was in the afternoon at a job site of a rock quarry. It was on a weekend and I was working there by myself.

I saw in the distance coming toward me a small plane or a glider. It was probably 200 feet high. I really did not give it any thought, but did notice it more as it was coming directly toward my position. I was at the time operating a large earth-moving machine in a limestone rock mine.

The machine is a dragline and has a large 300 foot boom on it. As I was dumping the material from the bucket, I noticed this object since it was rather low and could possibly come in contact with the boom.

Where I was dumping the material, there was already a row of piles of material. Probably 20 piles in a row between my position and this object. Each pile was more than 100 feet high. I would dump the material at the top of the end pile and could see this thing in the background. I would then swing away from the pile and dig another bucket of rock and again return to the top of the pile.

I could see this object coming closer with each pass. When it was only one or two piles away from the pile I was dumping on, I stopped as it was going to hit the boom maybe.

I fully believed it to be a small plane. The large window in front of me was wide open and I could hear no sound from a plane. As I watched, to my total amazement, it did a u-turn right in front of the boom and above it. Once it had completed the turn, it flapped its huge wings slowly several times and then continued its glide back down the row of piles.

I knew that I was now viewing a living creature and not a plane. I stood there and watched until it was out of my sight. I have never seen or ever heard of any kind of bird of this great size. I have seen large turkey buzzards and eagles before. This was very large and was the size of an ultra light airplane or a Piper Cub.

I could tell that it was gliding and using the updraft that was coming across the lake where I was digging. The wind was coming across the water and then up the sides of the piles.

I asked different people that I knew there at work who had been around for many years if they had ever seen something like a huge bird. They laughed and thought that I was joking, so I keep it to myself (p. 13).

BIG BIRD

Griff (2008) shared his TWIDDER experience on a paranormal website, too.

In November, 2004, I was sitting in heavy traffic in a suburb just outside Pittsburgh, Pennsylvania. It was about 4 P.M. and I along with a ton of other commuters had been sitting for nearly an hour due to construction.

Now before I get to my odd experience, I would like to state that I am a grounded individual—a 'see it before I believe it' type person. Granted, I have observed oddities before but not on such a large scale with so many other witnesses.

I was sitting in my car with the engine off. Most people had their engines off, so it was relatively quiet other than the chatter of people conversing from car to car.

I spotted something in the air and instantly thought it was a plane—until it flapped its wings! I thought I was seeing things at first, so I kept watching it. Sure enough, it flapped its wings again!

It was a HUGE black bird with a 20-25 foot (possibly larger) wing span. It was easily longer and wider than a

112

bus! I took a second to look around and make sure that I wasn't the only one who saw the thing.

Everyone around watched it, jaws agape, stunned. People were recording it with their cell phones. The giant made no effort to fly off. It calmly stayed in sight for a good five to ten minutes before it flew west and faded from sight.

I have seen large birds before, everything from eagles to condors, but this bird dwarfed all of them. It resembled one of the big birds from those *Lord of the Rings* movies, and I felt like a tiny hobbit watching it! I still consider myself a skeptic, but I, as well as hundreds of other people became believers . . . (p. 17).

PAST FOR ONE, FUTURE FOR THE OTHER

On June 2009 Phil T. wrote about his 1978 summer TWIDDER experience.

In the summer of 1978, I was working as a night desk clerk at the Grand Canyon, Arizona. I should say at the outset that the Grand Canyon does strange things to time; go there and you might experience it yourself. We were assigned quarters—similar to a college dorm—as part of our pay package.

One afternoon my roommate, taking great pains not to wake me, had entered the room to use the bathroom. Out of the corner of my eye I could see him, but standing at the foot of my bed was another man who was staring at me with a look of complete astonishment.

Thinking he was a new employee who might have gotten misdirected or had been assigned this room in error by personnel (unfortunately, not an uncommon occurrence back then), I rose up on my elbows, squinted (near-sighted) to ask him if he was lost.

He vanished completely!

I asked my roommate if he had seen anything. He hadn't. Now this is in broad daylight. I concluded that he was from the future, I was from the past, and we vanished in front of each other. I wondered if he wondered just who was sleeping in his bed.

A TRIP FORWARD FOR AN 1800'S SOLDIER

Dan (2009) and his family had a TWIDDER experience while they were on vacation in 1992 at the sight of the Battle of the Little Big Horn in eastern Montana. Also known as Custer's Last Stand, the battle was an armed engagement between a Lakota–Northern Cheyenne combined force and the 7th Cavalry Regiment of the United States Army. It occurred on June 25th and 26th, 1876, near the Little Bighorn River. And so Dan's story begins.

> I have a great-great uncle named Myles Keogh who rode with Custer and was the Commander of Company 'I.' I wanted to see just where he died.

> I had to explain to the Rangers who I was and how I was related to Myles, and after I told them in detail my relationship, they allowed me and only me to go to where he died and do a rubbing on his stone.

> It was about 105 degrees when we drove out to Reno-Benteen (Ridge) Battlefield. We were the only ones foolish enough to go out there in that heat, and there was not a soul in sight.

When we parked and got out and started to snoop around, a soldier in full uniform came up over the ridge from the direction of the river!

We started talking with him about the battle, and the kids were asking questions. He looked wiped (bushed/tired) so I went to the cooler and gave him a cold soda. He looked at it very strangely but didn't open it.

We explained our relationship to Myles and the soldier started telling us things that only a person who knew Myles could know. For example, Myles was a spy/informer/assassin for President Grant. Grant hated Custer for a slight at the end of the Civil War, and Custer was going to run for President after the battle (common knowledge).

Then the soldier told us how Uncle Myles was loved by his men because he shared his liquor. This was something I always wondered about, but I didn't want to ask him if Uncle Myles was a mean drunk. Myles had numerous casks of whiskey with the pack train.

Here's how Custer was shot as he tried to cross the Little Big Horn River in a flanking maneuver. The soldier told us that Uncle Myles was a crack shot with a 44-40 rifle. After Custer's battalion was pushed up onto Custer Hill by the Indians, Myles tried to save his own company, but before he left, and knowing it was a futile effort to escape, Myles put a round into Custer's head for getting all of his troopers killed.

Suddenly the soldier said that he had to get along, and he turned and went back over the ridge and out of sight. We went to the car. We thought this was pretty interesting but didn't give it too much thought.

But then I started to wonder how he could have known all this because everyone that went with Custer was dead, and the story couldn't be told unless that soldier was there himself!

I'll never forget the soldier. Honestly, the air seemed charged with electricity when he came into view.

> His eyes were piercing, like he was looking right through me.
> His voice sounded hollow and distant, with a slight drawl.
> He was a sergeant; his uniform was dusty-dirty and well worn.
> He wore a kepi.
> He wasn't clean-shaven, but he wasn't bearded. He looked weathered and raw.
> He seemed awestruck by my wife and kids like he had a wife and kids and missed them very much. (If he was a re-enactor he could have won an Oscar with his performance.)
> He never acted or sounded like a Ranger in a 'living history' mode; he just referred to each area as "over there," not by its common National Memorial name.
> And he didn't point with his finger, but just like a Native American, he pointed with his nose, throwing his head back and using his nose as a pointer.

That area around Reno-Benteen and the path to Uncle Myles' marker stone was the most electric, spirit-filled place I've ever been. I could almost hear the war cries of the Sioux warriors. Not spooky, but—this'll sound stupid or crazy—almost like the old Twilight Zone episode about the

National Guardsmen who went back to the time of the Battle of the Little Big Horn.

I am a very believable Civil War re-enactor, but that soldier never strayed out of character. I've never see anything as authentic as he was. That's something me, the wife and the kids still talk about.

Chapter 5.

Fast Forward

Sir Isaac Newton believed there was nothing special about the speed of light; if you raced along a light beam, it remained stationary. A young Albert Einstein didn't think Newton's view made sense.

As a college student, Einstein discovered that you can never race alongside a light beam; it always moves away from you, no matter how swift you are.

In fact, depending on how fast you move in relation to light, time will beat at different rates. The faster you travel, the slower time ticks.

Should you achieve the speed of light (186,287 miles per second), time ceases totally. On our plane of existence, time is NOT fixed at all; it is relative.

GPS systems must take this into account, or they'd be about as much navigational help as my maternal grandmother. (We'd ask her which way to turn, and she'd wave her hand like a pendulum bob, announcing, "That way, that way!")

In Einstein's general theory of relativity, space-time is a fabric that can stretch and shrink. Under certain circumstances the fabric may stretch faster than the speed of light.

Is that what happens when a TWIDDER results in a fast-forward phenomenon? Has the experiencer, like a spider on a wind-stretched web, been taken for a short ride on a briefly distended thread of space-time?

To quote T. S. Eliot, "The journey, not the arrival, matters." In hindsight, all of the country roads down which my grandmother's ambiguous directions took us made the journeys that much more memorable. But NEVER quicker.

STRANGE THINGS

Charles Lindbergh (1902-1974) was an American aviator, author, inventor, and explorer.

On May 20–21, 1927, Lindbergh, then a 25-year old U.S. Air Mail pilot, emerged from virtual obscurity to almost instantaneous world fame as the result of his solo non-stop flight from Roosevelt Field on Long Island to Le Bourget Field in Paris in the single-seat, single-engine monoplane dubbed "Spirit of St. Louis."

Lindbergh, an Army reserve officer, was also awarded the nation's highest military decoration, the Medal of Honor, for his historic achievement.

In his final book, *Autobiography of Values*, Lindbergh records a flight that defied description. It took place in 1928.

He departed Havana, Cuba, at 1:35 A.M. on a long-planned flight direct to St. Louis, Missouri. His itinerary called for crossing the Straits of Florida and then heading straight to St. Louis. Most of the high flying would be dead reckoning. (The navigation aids we take for granted today weren't even wishful thinking in the 1920's.)

As he left Cuba behind and flew over open water, strange things began to happen. First, his magnetic compass went nuts, spinning uselessly.

However, his aircraft also had an earth induction indicator, the primary (although by today's standards primitive) piece of navigation equipment of the time. But the EII went bananas, wandering erratically.

Lindbergh had no way to check his heading or to use timing and dead reckoning to figure his position. He kept flying as steadily as he could, using star references to hold course. Then the heavens went bonkers! In an instant, a blinding fog formed around pilot and plane.

Now not-so Lucky Lindy swooped down and flew as low as he dared in the dark. Then even the air turned against him and became turbulent.

Dawn finally streaked the eastern horizon. Lindbergh couldn't believe the sky. He described it as looking like "dark milk."

The sun finally prevailed enough that he could see the surface below. He was over land! His compass started working, and only a moment later, the EII did as well. Perfectly.

It took a while for Lindbergh to locate his position. For several minutes he scanned the shoreline features and then his charts. If they matched, as they seemed to, then he was far off course. So far that it exceeded any distance he could possibly have reached with the fuel on board in the time since he left Cuba. And match they did.

Lindbergh had experienced a time-slip.

300 MILES FROM HOME

On November 2004 Cher wrote on a paranormal website about a scary drive to town one night.

I was leaving my driveway to drive one mile to town. There were woods and a few houses along the way.

I had just pulled out and there was a thick fog rolling in. Very creepy at night. I had only gone a half a block at most, very slowly through the fog, when I saw a bright light. This was very strange since there was nothing near my house.

As I drove closer, I saw a Phillips 66 gas station. It wasn't there the day before! I pulled into it in shock and a man came out. I asked him how a gas station could have been built so fast, and he looked at me strangely. He said it had been there for eight years.

I asked him where I was, and when he told me, I began to shake uncontrollably. I was 300 miles from home! I looked at the clock; I had left my driveway only TWO MINUTES before. It took all night to drive home, and I shook most of the way (p. 16).

HIGHWAY 13

Corrine Kenner and Craig Miller, editors of *Strange but True* (October 1997), tell the story of John Gerzabek of Phoenix, Arizona, who had an "other-worldly" experience on the road from El Paso to Phoenix in March 1994.

In 1994 I drove a truck for a Phoenix-based company. One night I was pulling a load of minivans from Louisiana to Phoenix in an 11-ton Chevy with a 48-foot trailer.

The truck was really acting up. It just seemed to be fighting me and falling apart. I didn't think it would get to Phoenix. When I got to El Paso, I stumbled into the Petrol Truck Stop. I was tired, my head hurt, and my bad arm was throbbing.

I drank some coffee and talked with some of the other truckers. I knew that I should rest, but I was too anxious about getting home. When I left the truck stop I checked my odometer, recorded the reading, and took off. I saw an 'Exit Three' sign for Highway 13. I took it.

I couldn't believe my luck. There was no traffic at all. The road was like a clear white ribbon. The pavement was smooth and even, my truck was running smoothly, almost like new. There was no scenery, no road markers.

All I passed were three old Indian cemeteries. I figured that I had at least a seven-hour trip ahead of me, so I was glad that I had a good road and my rig was doing all right.

[At one point] I looked out the window and there to the left, sitting quietly on top of a flat rock, was a huge pure white wolf. It made no effort to move but stared at me with piercing green eyes.

I had barely passed the wolf when I came across my first road sign. I couldn't believe what it said: 'Arizona state line, nine miles.' I had been on the road for less than two hours!

Soon I came across the only exit I had seen off Highway 13. A sign read, 'Phoenix, 18 miles.' I pulled into my shop exactly two and a half hours after leaving El Paso.

The trip usually takes seven hours. By the time I went up 35th Avenue in Phoenix, the truck was acting up again. My mechanic did not understand how I had driven it 412 miles, and neither did I, but I will always be grateful for Highway 13 (pp. 29-30).

160 MILES IN AN HOUR

Sheri Lowe (2007) and a friend had a fast-forward time experience in the summer of 1986. Their journey started in Saskatoon, Saskatchewan.

It was the beginning of August and was one of the most beautiful summers of my memory. Each day had been idyllic. Hot, but not killer hot, and lush, which is not typical for our hot dry prairies.

The lushness was due to beautiful thunderstorms occurring almost every evening, which cooled the heat of the day, invigorating the night with a magical aliveness. You'd feel the storm approaching by the way hairs on your arms would rise, and the perfumed air held the hint of ozone. The storms were spectacular but not dangerous, and this too is atypical for this land of extremes.

My boyfriend was an American who[m] I had met when he came to Canada to reacquaint with his son from a previous marriage. To make a long story short, he was unable to stay in Canada and get a working visa, so he had to go back to the U.S.

Naturally, I was heartbroken. He decided he'd return to Minneapolis and continue to pursue a working visa from there.

I decided I would take a month's vacation, drive him to Minneapolis, and help get him settled. From Saskatoon we planned to drive to Regina, Saskatchewan, which is 160 miles away, refuel, have a stretch, then head to Minot, North Dakota, stop for the night, then head to Minneapolis.

Michael and I were both exhausted from months of fighting the deportation order and depressed because he lost his case. It hit Michael hardest having to leave both his son and me. The stress gave Michael a nasty case of walking pneumonia, so as I drove, he slept.

I distinctly remember checking the time before we left because I was bugged we were leaving so late. It meant at some point I'd be driving in the dark, which I disliked.

As usual, I had the radio on and heard the DJ confirm the time during his hourly newsbreak. I remember making a quick mental calculation that we should reach Regina between 6 and 6:30 P.M. And that I'd have light until about 10 P.M.

We were only 30 or so miles out of Saskatoon on the Trans-Canada #1 Highway when I started feeling odd. My physical experience inside the car began to change.

Everything quieted, sort of muted—even the radio—like being wrapped in a wad of cotton. I looked over at Michael to comment, but he was sleeping soundly.

Then I noticed that the road noise and bumps one experiences when traveling in a vehicle had disappeared. The car felt as if it were hovering a few inches above the road and was sort of sailing through the air.

I saw no other cars on this very busy highway, which also struck me as odd. I distinctly remember feeling everything was syrupy. The closest thing I can compare it to was when I'd lane swim a mile and be so exhausted until I'd break through it and go into a state referred to as a 'runner's high,' this whooshy, slow motion feeling.

When we got to Regina, I went to refuel and my first shock came when I noticed we had used almost no gas at all. My second shock was when I asked the pump attendant for the time. He said it was 5 P.M.!

Even though I already knew the answer, I still felt compelled to ask the attendant if Regina was on the same Central Standard time system as Saskatoon. He looked at me like I was the dumbest blonde he'd ever seen. By this time Mike was awake and he, too, was as amazed and baffled as me.

It was physically impossible for me to have made 160 miles in one hour—even if I had been speeding! What's more is the very fact that I knew something weird had been going on; it had been physical.

The rest of the trip was normal, but I will never forget that experience (p. 20).

TEN MINUTES
J.H. (2003) had a similar experience of missing time. Before retiring in 1998, J. H. had to travel 30 miles every day to and from work. The route was very isolated, the road a narrow ribbon through woods.

I had specific landmarks that I used to clock my travels. They were about five minutes apart.

On more than 30 occasions, I would leave my home town and five minutes into what would normally be a 30 minute drive, I would be five miles out of the town I worked in.

According to the dash clock in my truck, I made a 30 minute drive in about 10 minutes with no memory of the missing 20 miles or the time I should have gone through.

This has never been mentioned to anyone outside of my family due to my career in law enforcement (p. 25).

THE DRAGON'S TRIANGLE
BellaOnline by Deena Budd (n.d.) narrates an experience centered on a triangular area in the Pacific. In recent years much has been written about the Bermuda Triangle and the mysterious phenomena associated with it. The Bermuda Triangle is a triangular shaped area of the Atlantic Ocean stretching from the Straits of Florida, north-east to Bermuda, south to the Lesser Antilles, and then back to Florida.
   On the other side of the world, there exists a lesser known but similar area of ocean known as the Dragon's Triangle. The

Dragon's Triangle follows a line from Western Japan, north of Tokyo, to a point in the Pacific at an approximate latitude of 145 degrees east. It turns west south west, past the Bonin Islands, then down to Guam and Yap, and west towards Taiwan before heading back to Japan in a north northeasterly direction.

It was within the Dragon's Triangle that Amelia Earhart's plane was lost. Another pilot, entertainer Arthur Godfrey, experienced a TWIDDER in the same area.

Godfrey was an American radio and 1950's television broadcaster. He was known for his down-home, folksy demeanor and ukulele playing. I'm old enough to remember my father singing Godfrey's 1947 hit song, "Too Fat Polka."

Godfrey learned to fly in the 1930's and remained an avid aviator for the rest of his life.

In the late 1950s, Godfrey was flying over the Dragon's Triangle when his instruments failed. Flying blind for an hour before his instruments returned to life, he made it to Tokyo and found that he had 'lost' thirty minutes.

SAME PLACE, TWO DECADES LATER

Steven Johnson, in his Mercury Rapids article (n.d.) shares the tale of Lieutenant Colonel Frank Hopkins, an advisor to the 106th Air Transport Group. Hopkins was flying in a C-97 Stratofreighter over the Dragon's Triangle in 1968. He was a navigator and, as protocol required, he used star navigation to plot their course every hour.

Three hours into the flight, he took another celestial fix and was astonished to find that their position was more than 340 nautical miles down from their intended course.

On landing, he told his duty officer that the aircraft had been "dropped" many hundreds of miles ahead of their plotted course. Hopkins maintained that the area was prone to peculiar forces that posed danger to planes and ships that ventured across it.

FAST FORWARD IN A GEO METRO (I know, sounds like an oxymoron.)

Don and his wife had their TWIDDER in the Nevada desert in 1997. They were driving their small, three-cylinder Geo Metro to Laughlin, Nevada, from San Diego, California.

"We took the back way to Laughlin on 8 East through the desert to 95 North," says Don. "Around dusk, our windshield had a rather large collection of bugs, so we pulled into the only gas station at Vidal Junction to clean the glass."

As they pulled into the mini-mart station, Don noticed a seedy-looking character at the pumps staring at him and his wife. He had greasy hair, a disco shirt and a leather vest, and was driving a beat up, early '70's Toronado.

Don quickly cleaned his windows and got back into his car. The greasy dude was still staring at them. Darkness was quickly approaching as Don pulled back onto 95N, hoping the creepy stranger would not follow them. But he did.

"Immediately, I had a bad feeling," says Don, "Knowing that in the lonely desert there are bandits who prey on tourists, nudging their car, then robbing them after they pull over for a fender bender."

Don knew his Geo Metro was no match for the stranger's aging muscle car but tried to outrun him anyway. Even managing to briefly reach 80 mph with the gutless Geo, the stranger stayed on his tail.

Don and his wife began to get frightened. "I told my wife to get the gun out of our backpack we travel with for protection."

But then something weird happened. "A split second later, with no turn offs, we were suddenly alone on the highway," Don remembers.

No greasy dude, no one but oncoming traffic miles ahead.

A few minutes later we were at 95N and Hwy 40 in Needles. But it was 7:00 P.M. when we left Vidal Junction, and Interstate 40 is 55 miles north of there.

Now it was only 7:20 P.M. Somehow we'd made the entire 55 miles in twenty minutes! No Geo Metro can fly like that. We would have had to go at least 130 mph or more on a two-lane that twists through some mountains.

The other strange thing was we had a strange floating feeling just as the other car disappeared.

FROM FITZROY FALLS AT WARP SPEED
Aussie Joe (2006) wrote about a fast forward experience from September 2003.

I was taking a drive with my girlfriend Julie in the south coast region of Sydney. We had left Dapto and proceeded to drive up Maquarie Pass, found at the base of Albion Park New South Wales; a pioneer road leading up the escarpment to Robertson, built in the 1880's. Once at Robertson we followed the signs to Fitzroy Falls and arrived at our destination in two hours total and had a great visit.

On our trip back, following the only known route from Fitzroy Falls back to the Robertson and Macquarie Pass, we drove the quiet straight road.

After about 35 minutes, Julie and I discussed how we never recalled that the road descended and swerved and [wondered] where a storm came from nowhere all of a sudden.

We simply thought it was a small formation that usually occurs over the escarpment, but it was weird. We admired it at first; the clouds were thick, very dark formations that seemed to be inverted clouds.

It rained very hard and had loud thunder and lightning flashes. It stopped as abruptly as it had started, but it was still overcast with crazy cloud formations that seemed very low to us.

Three minutes later, we arrived in a town call Berry, a town destination that would at least have taken four and a half hours to reach. Missing the town of Robertson altogether, our entire journey took only 45 minutes.

Yet we saw no turnoffs to Robertson, Macquarie Pass or main interchanges. As far as we knew, we had just left Fitzroy Falls just a short time ago. We had no explanation of how we arrived at Berry, a town we didn't even want to go to, and how we had seemed to jump time and distance.

We weren't drinking, and we don't do drugs. We did debate and puzzle over this, trying to reach a logical explanation . . . and that storm? What was that? It was as if it were a collection of storms' greatest hits!

We agreed to just agree to stop debating our theories as it was frustrating each other, but we both agreed something didn't add up.

How could we arrive at an unplanned destination ahead of the designated time it would have taken to get there? I am eager to know of anyone that had a similar experience in taking this route on the escarpment (p. 2).

ARRIVING THE SAME TIME HE LEFT

"Nettdan" (2001) experienced a TWIDDER that occurred in 1986.

I had left home at 5:45 A.M. (I checked every clock in the house since it was the first week on the job and I didn't want to be late.)

This gave me a 15 minute window to walk to the train station to catch the 6 A.M. to Grand Central, Adelaid, South Australia, which I caught, no worries.

I remember freezing while I waited. I would travel each day to Grand Central, about a 50 minute journey, and just have time to catch a connecting train to the suburb where I worked.

After getting on board, I huddled up to keep warm and fell asleep. I woke as the train pulled into Grand Central and my watch said 6:54 A.M., which gave me about four minutes to catch my connection.

Rushing through the station, I ran into my brother with whom I worked. He was having a very large and leisurely breakfast.

I told him to get moving [because] we had three minutes left. He looked at me like I was nuts and said we had ages since he had apparently missed his alarm and arrived quite early.

Looking at my watch, it said 6:57 A.M. He pointed out it was obviously wrong and that the station clock said 6:05 A.M.

Weirder still, it was right. We asked several other passengers who all confirmed 6:05 A.M. How do you travel 50 minutes and arrive at the same time you left? Note: The clock at my home station had said 6:04 A.M. as I got on the train (p. 13).

THE TIME JUMP

Martin Caidin, in his book *Ghosts of the Air: True Stories of Aerial Hauntings* (1994) shared what was to have been "just a jump across some water" for pilot John Hawke. Instead, it turned out to be a time jump. He was scheduled to fly from Fort Lauderdale to Bermuda, a route he'd flown, literally, hundreds of times. He was flying an Aztec with long-range tanks direct from Fort Lauderdale Executive Airport in Florida to Bermuda. The forecast was for puffy cumulus clouds between four and eight thousand feet.

But what happened on this flight was something I'd never before encountered. I was on autopilot, everything as neat as a pin. It was like being at home. Everything was perfect until I found myself staring at the magnetic compass. I was staring at it, all right, but I couldn't see the stupid thing.

Oh, the compass was still there. But that compass card, that idiot thing, was spinning so fast, it was a blur. And I began to feel as if I were passing out. Like slipping under an anesthetic.

I put my head back on the rest and had a good look at the sky. It wasn't there. Just creamy yellow *everywhere*. No clouds, no water, no horizon, no blue. Just yellow.

Fifty-nine minutes later, still flying northeast, I could now see the sun. But I didn't know where I was. I felt fine but drained.

I looked up, and there was a lovely contrail. Just beautiful. Got on the horn right off and gave them a call. They came back. I told them I was under them and asked where in the devil I was.

I didn't believe what they told me. Not at first. I couldn't believe it. I was seven hundred miles from where I'd been just an hour before. I'm in a bloody machine cut back to economy cruise, and oh, a ground speed of 180 or so. And in that hour the bird covers a distance of seven hundred miles.

It's impossible, of course. But it was true. I turned due west and landed in Virginia (pp. 228-229).

INHOUSE WEIRDNESS

In July 2006, Luci Stravato explained her experience with missing time. She said that it might not have been as spectacular or frightening as some, but "I guarantee that it is 100 percent true, and it has baffled my husband and me for years."

Dave set the alarm for our usual 7:00 A.M. and off to bed we went. Another ordinary evening. Early the next morning we awoke to the sound of my lab/boxer, Bear, barking her head off. She usually only does this when there is someone or something (like a neighborhood cat) lurking around the house. Dave and I got up to investigate, but before I left the room, I noted the time: 5:30 A.M. Cool, I thought, another hour and a half of sleep.

We both walked to the kitchen window, for it was the focus of Bear's boisterous attention, looked out and confirmed that no one was outside. Must have been that darn cat.

We both grabbed a quick drink from the fridge and padded off back to bed. Dave had to make a pit stop, but I was more than ready to grab that hour and a half of sleep.

I jumped into bed and was just pulling the covers over myself when I noticed the time: 6:30 A.M. WHAT?! Well, although I was sure about the time, maybe I was bleary-eyed when I jumped up earlier, so I called to Dave.

'Honey, did you notice the time when we got up?'

'Yea, he called back, 'about 5:30. We have a little longer to sleep.'

Okay, so now I'm officially freaking out. It took an HOUR to walk to the kitchen and back? What happened to us? I have no explanation and no information except I promise you it was 5:30 when I woke up and 6:30 when I came back from the kitchen. Weird, huh?" (p. 6)

WHIRLWIND TRIP
Vicki (2006) told a story of a possible wormhole on Earth. One weekend Vicki and her 15 year old daughter planned to head out to visit her older daughter at college.

My 15 year old had spent the night at her father's house, 12 miles away (it takes around half an hour to get there), so I had to first pick her up before we started out.

When I arrived at my ex-husband's, she was not yet ready (of course), so I went in to talk for a few minutes. From her father's house, it usually takes two and a half hours to get to my older daughter's college.

While driving, there was one area where I remember that we were so engrossed in conversation that I didn't remember seeing anything of the area. We didn't realize until we got to the college what the time was and that I had traveled 150 miles in an hour and 15 minutes!

This also included the time I spent at my ex-husband's house when I went to pick up my daughter. We used little or no gas for this trip, as when we arrived the tank was still on full. I have a hybrid car, but I don't think it is supposed to get that kind of mileage (p. 24)!

CONDENSED TIME AT THE MOVIES

One afternoon in the late '80's "Retro" decided to see a movie by himself. He told this story in March 2005.

I arrived to view the movie at 4:45. I parked the car, ran to the outside ticket booth, and paid for my ticket.

I walked in, watched the entire movie, and when the credits rolled, I got up and walked out front. I felt funny when walking through the theatre door to the lobby. It felt like I was standing in more than one spot at the same time. It didn't hurt and it wasn't uncomfortable, just unusual.

I thought maybe it was because I had been sitting for quite awhile. On my way out of the theatre, I noticed I didn't see anyone. Not one person in the actual theatre lobby. I walked outside, which seemed a little bright for the time of day it should have been, and got into my car.

After starting the car, I noticed the time showing was a little after 5 P.M. I looked at my watch, which showed what I thought was the correct time of 6:30 P.M.

After walking into my home, I noticed that the clock had 5:15 P.M. on it. I don't know what happened that day. I know I saw the entire movie, but apparently I was only in the theatre for 15 minutes. It still haunts me to this day (p. 23).

ODDNESS IN KANSAS
In July 1962, Bill (2007) had an odd experience in Salina, Kansas, that he still befuddles him.

I traveled from point A to point B in an extremely short amount of time.

It was a very hot summer day and I had just called my girlfriend (my wife now of 43 years). I told her that I would walk down to her house to see her.

She asked me not to because it was too hot. I told her that I would be okay. I estimated that it would be a walk of about 30 to 45 minutes.

I stepped out of the front door of my home, and INSTANTANEOUSLY I was standing in front of her house.

I knocked on the door and she answered. She insisted that I must be playing a joke on her and had called her from one of the neighbor's because she'd just hung up the phone, and just walked across the living room to answer the door.

I did not know any of her neighbors, and cell phones didn't exist at that time. This had never happened to me before, and never again since. I have no way to prove this, but it is a real experience to me (p. 2).

COUNTRY ROAD

In the late 1990's, Aaron Berke was a fifth grader in Oregon when he had a timeslip experience that he shared in April 2008. His family was living in the Colton/Beavercreek area, about 30 minutes southeast of Portland.

It's a very rural area. We lived in Beavercreek, population 5,000. I went to elementary school in Colton, about 15 minutes away. [There was] only one way to get there along rural roads.

To the best of my memory, it was 7:00 P.M. after a basketball practice at my elementary school. It was dark and we had been driving for about two or three minutes of the drive home, which my mother and I had driven many, many times.

We turned onto the more main road, which we would drive on for about eight to ten minutes, at which point we would turn off and be at our house in three or four more minutes.

Only about 30 seconds after we turned onto the main road, my mother and I were talking moderately, and she suddenly braked to a slow pace and said, 'Aaron, look!' And as I looked out at the road, it was absolutely black: no road lines, no edges of the road almost as if we had just driven off the road into the woods.

She continued at about five to ten miles per hour for just about 15 seconds when suddenly the road became visible again, lines, edges and all.

We proceeded to talk about how weird this was when we came upon a very small town-like area. Confused, we then realized by turning around that we had driven ten minutes past the turn-off to our street.

This means we had traveled a 20 minute distance in less than two minutes. It was as if we turned onto the road, blacked out for one minute, then appeared ten miles further than we should have been. We were somewhat stunned and joked about returning home to our family being years older.

This experience is the only paranormal one I have ever experienced in my life. And I swear by it; no one could convince me differently, or my mother. There was some sort of a wormhole on that country road (p.12).

SECONDS TO SYRACUSE

On September 2009 Robert Asher told of a 1984 incident in his life that has stayed with him for years. Asher was driving from New York City to Syracuse, New York, looking for a house.

A realtor had called us because he had found a home for us to look at. At the time, I was working at the Brooklyn Naval Yards and I received the call from my wife telling me we had to go to Syracuse that night.

It was autumn. I got off work at 6 P.M. [and] went home to pick up my two daughters (at the time they were young girls). We left the house about 6:30 P.M. and arrived at the New Jersey Turnpike at about 7 P.M. We stopped at a rest stop and bought coffee and snacks.

Now mind you, this was not our first trip to upstate New York, so we knew where we were going and how long it would take—about five hours.

When I got back to the car with the snacks, it was pouring rain; we could barely see in front of us. There was a faint glow on the highway just beyond a hill, and my wife turned to me and said, 'It looks like we're driving off the edge of the Earth.'

I turned to look at her and when I turned back to the road, it was daylight, sunny, and *we were on the outskirts of Syracuse!* That was a total distance of about 300 miles, and eight hours had transpired!

I turned back to my wife and woke her up. The children were still asleep in the back. I told her where we were and what time it was, and she said that there was no way—she had just gone to sleep a minute ago.

I can only remember getting into the car at the rest stop in New Jersey . . . and I ended up in Syracuse. I couldn't remember driving to Syracuse.

I know I didn't fall asleep at the wheel and drive for 300 miles. And I didn't pull over to sleep on the road because I had my kids in the car.

It could not have been driving hypnosis because I had to drive through the Shawungunk Mountains. I am 62 years old and am still trying to find an answer to my mystery.

HOW TO SKIP A TOLL BOOTH—THE WEIRD WAY!

In August 2008 Katie wrote about her experience in gaining time.

The year was 1992. I was attending school in Davie, Florida. I lived in Lantana, Florida, at the time. The trip from Davie back home to Lantana took about an hour.

One morning my classes had been cancelled, so I headed back home. My route home was always the turnpike to the Sample Road exit (I had to go through two toll booths) over to I-95 and on north to Lantana.

Well, it was a nice, sunny, south Florida day when I left campus. I had my toll money out in the tray next to me (I always liked to have it ready ahead of time so that I wouldn't have to dig through my purse).

I was listening to the Lee Fowler talk show on the radio (sadly, Lee has since passed away) and just sort of zoning out the way you do when you've made the same trip a hundred times.

I went through my first toll booth, everything normal, when I saw the exit sign for Sample Road coming up. I was startled to see it, and glanced at the clock in my car and at my watch.

They both confirmed that I had only left school 10 minutes before. I had somehow made a 30 minute trip in 10 minutes! The money for the second toll was still in the tray beside me. I exited Sample Road and pulled over. I was confused and upset and half expecting the Highway Patrol to be after me for running a toll!

I calmed down and knew that there was no way I could have crashed through a toll booth gate and not noticed it! How to explain the time? I couldn't. The radio worked the whole time, everything seemed normal. Except I somehow made a 30 minute trip in 10 and somehow skipped a toll booth (p. 26)!

A BRIEF LEAP IN TIME

In July of 2004, Sue wrote of her TWIDDER that had occurred in 1994. She and her boyfriend Jim were driving to their home in Fallbrook, California.

There are two roads into Fallbrook from the north and from the south. We approached from the south heading north on Mission Roa, which is a four-mile, curvy, two-lane road.

It was a Friday evening and I blurted out how weird it was that we hadn't passed any cars in the opposite direction since our turn onto Mission Road. Fallbrook is a small town, but this was a very well-traveled road.

I knew it was 6:24 P.M. because I had just looked at the big amber display on the stereo when all of a sudden the car died, the pedal went hard and the lights went off.

I remember feeling confused, thinking that we ran out of gas, yet knowing that we had plenty of fuel. We coasted off to a dirt turnout immediately ahead.

Jim asked what happened and I had no answer, but I felt like I had just dozed off or had gone through Jello or . . . I don't know what.

I put it in gear, Jim jumped out, came around to my side and opened the door. I jumped into the passenger seat, he got in, cranked it over and—vroooom!—off we went.

As we pulled away I felt irritated, a little nervous, but sure something really weird had just happened. Then I glanced at the clock on the stereo again. It was 6:36 P.M. Twelve minutes had passed when, being generous, the whole incident couldn't have taken more than one and a half minutes, max (p. 16).

EARLY TO WORK

In February of 2009 Harland G. told the story of a phantom ambulance. Harland arrived at work much earlier than he expected one morning in July, 2008.

I was on my way to work around about 6:30 A.M. It was really clear that morning. I was about 10 minutes away from where I worked when I drove into some fog.

I thought it was a little weird because there was no fog anywhere else and there wasn't a river or anything to cause fog in that area.

When I drove into the fog, it was maybe about a minute in when I saw lights flashing through the fog. As they got closer, I could see that it was maybe an ambulance.

I slowed down a little, waiting on it to pass. It seemed to take a while for it to get to me. I thought it might have stopped. Then in a few seconds the ambulance appeared in front of me.

When it started to pass me, it seemed to move in slow-motion, and while it was passing, it never made a sound. No sounds of the engine or any sounds on the road. There was nothing.

After it passed, I rode out of the fog and arrived right in front of the parking lot of my job. I was 10 minutes away from my job, and I got there in two! I don't know where that time went, and I don't know what happened in the fog.

Chapter 6.

So Slow

For twenty-five years I had a horse whose nickname was Go So Slow. He may have been half hot-blooded Arabian, but it was his quarter horse dam he took after; conservation of energy at all costs.

Ask him to go faster than a leisurely walk, and you'd better be prepared for a highly entertaining bucking session. On the other hand, he could hear a bucket of oats rattle from half a mile away. At which point he'd go to Warp One in 0.3 seconds, and the rider would find herself whiplashed and in need of a change of underwear.

While we were able to put cause and effect to Go So Slow's fast forward and so slow phases, TWIDDERS leave us flummoxed by these experiences.

If a fast forward event is caused by a stretch in space-time, is a so slow one then caused by a relaxing of space-time fabric, a wrinkle in time? And, in some instances, is it more a matter of lost time than sluggish time?

PHYSICS CLASS

An experiencer who chooses to remain anonymous wrote about a So Slow TWIDDER he/she lived through in his/her senior year in high school.

Keep in mind that even though the description is long, the entire incident in reality probably lasted less than 10 seconds.

Anyway, this happened in my physics class. The teacher was late, so we were all (about 25 students or so) sitting around goofing off, as teenagers are apt to do.

One of the guys, Kevin, was up in the front of the room writing something on the blackboard. The physics room was equipped with those heavy black lab desks, and there was one of those right in front of the blackboard. It was raised to mark it off as the teacher's desk (I guess).

We had done an experiment the day before with a big bowling ball (as opposed to a candlepin . . . this was in New England, by the way), and the bowling ball was still sitting on the desk where we had left it the day before.

All of a sudden, for no apparent reason, the bowling ball started slowly rolling off the desk. It started in the middle of the desk and was rolling to the right if you were facing the desk from the back of the room.

Now here is where things get weird. When I say the ball was rolling slowly, that is an understatement. The ball was in slow motion.

Kevin was still up behind the desk at the blackboard, and he was just standing there with this totally spaced out expression on his face, watching the ball roll towards the edge of the desk.

Everyone else in the room was just sitting there, and we were all watching the bowling ball creep ever so slowly across the desk. I remember trying to say something to Kevin, but it was like my body wouldn't respond to my thoughts. Like my brain was moving in normal time, but my body was in slow motion.

146

Finally one of the guys in the class managed to say (in a very slow, deep voice), 'Kevin . . . the bowling ball . . .' Kevin was sort of standing sideways to the desk, and he whipped around (still in slow motion . . . imagine watching a Stallone movie where Sly is jumping or running in slow motion) and attempted to run after the bowling ball, which by this time was teetering at the very edge of the desk.

He took one or two very slow steps before the ball finally fell off the desk and crashed very loudly on the hard floor.

The crash of the ball seemed to break the spell that held the classroom, and suddenly everything was back to normal. Everyone was just sort of sitting there with wide eyes and open mouths, totally stunned. You have never heard a quieter room of 17 year olds in your life.

After a few seconds, everyone started giggling and talking nervously about what had just happened. Everyone agreed that it had been like time was slowed down, especially poor Kevin, who said he had been trying with all his strength to move more quickly and couldn't (he was a running back for the football team).

Has anyone else ever experienced anything like this? It was very, very weird. I would write it off to my imagination if the WHOLE class hadn't experienced this simultaneously. I guess it is worth noting that a couple girls in the back of the classroom didn't notice anything weird and thought we were all nuts. They, in fact, hardly noticed the whole incident until everyone else started freaking out.

NO EXTRA FUEL

Remember pilot John Hawke (in Martin Caidin's *Ghosts of the Air* book) and his fast forward TWIDDER from the previous chapter? Some years later he was flying a Riley Dove from Miami to Bermuda. Something happened again; only this time it was a So Slow phenomenon.

Once again, the horizon and ocean disappeared within a glassy yellow mist. Hawke kept flying, on . . . and . . . on. The normal time for the flight to Bermuda came and went, and they were still flying.

Hawke flew for another five hours, and then the world returned to normal. Bermuda lay dead ahead, and he went down for a normal landing.

On the ground he checked his fuel. The airplane had burned exactly the fuel Hawke had flight-planned before departing Lauderdale.

The only problem was that he had been in the air five hours longer than his flight plan called for, AND he had not burned one gallon of fuel over that required for the flight!

"That bloody machine gave us five hours of flying and, according to the fuel in the tanks, which I personally checked after landing, didn't consume a drop of fuel for that extra time" (p. 229).

BRUSHING AGAINST ANOTHER UNIVERSE

In August 2008 in Eugene, Oregon, Carol Ann (2008) experienced a So Slow TWIDDER.

> One day after work, I went to pick up my husband's prescription. Just as I entered the doorway to the store, everyone in the entryway started to move in slow motion, except one lady.
>
> I've never seen this woman before that I am aware of. She was average size and height, with past-the-shoulder-length blonde hair.

She was carrying something under her arm—a folder or notebook. She was moving quickly. There wasn't any room for me to get out of her way, so I was sure we were going to collide.

I leaned to the left, but she went right through my right shoulder, grazing the side of my leg. As she went through me, my shoulder was ice cold.

As I realized that we did not collide, I grabbed my arm, which was cold to the touch, and turned back to see her simply fade away.

I continued into the store, and it was almost like a memory lapse. The after-feeling was almost stranger than the moment of the incident. I felt that I had stepped into a different dimension for just a couple of seconds. I didn't feel that the woman was even aware of me. She was in a big hurry (p. 8)!

TAKING FOREVER

Kim and her husband had a weird experience near Laughlin, Nevada, in August 1999 that they shared years later on Stephen Wagner's Paranormal Phenomenon website.

"I was following my husband home from Las Vegas to Kingman, near Laughlin," said Kim. "He was on his Harley, and I was following in my car. We had made this trip several times before and knew exactly how long it took to reach home from the Hoover Dam: one and a half hours."

The weirdness began with their perception of the weather. And her story continues.

My husband swears it was raining in the distance and lightning was so close he could almost feel the electricity. I swear it was dry as a bone.

Also, I had a hard time keeping up [with him] as he was going very fast around the corners. I could see him in the distance, and suddenly there were a lot of cars ahead of me and behind him.

I thought that was odd since there was a mountain up one side and a sheer cliff down the other. There was nowhere for those cars to have come from. Oddly enough, as soon as I thought that was weird for them to be there, they were gone!

It seemed as though the trip was taking forever and I was getting really tired. When we arrived home, I thought it was really late and so did my husband.

We looked at the clock and the one and a half hour trip had taken *over four hours!* We are afraid of what happened to us during that missing time.

SLO MO SOCCER

In March of 1995 Molly wrote about a So Slow TWIDDER from her high school days.

Most of my friends are boys and in high school, we were all soccer crazy. One afternoon, we were all going for the school's soccer team tryouts. Our coach was there supervising.

After about two hours, he called for a break. During the break, Simon, my best friend, and I decided we wanted to kick a ball around just for practice.

So we were pretty much enjoying ourselves and our practice game got a little competitive. The ball kicked out of play, and Simon got to it first.

As Simon was about to kick the ball to the other end of the field, somehow, to me, he seemed to be moving in S-L-O-W—M-O-T-I-O-N.

I saw the ball suspended in mid-air. Somehow, I managed to reach out and pluck the ball right out of the air and just as suddenly, things went back to normal speed.

Simon kicked empty air, lost balance and fell backwards. Simon and I looked at each other blankly; then we heard some of our friends gasping and freaking out. That brought us back to our senses.

We looked over at the group of people; everyone was freaking out. Until that moment, we had thought it was just all in our over-stressed minds.

The coach, looking rather pale, came up to us and said, 'What the h--- was that???'

We didn't know how to respond. Then he continued, 'Simon, you looked like you were moving in slow motion, the ball . . . it was suspended in mid air and (looking at me) you . . . you seemed to be moving in fast forward speed . .'

We were all very shaken up by that experience. The coach actually postponed the rest of the tryouts until the next day. I have absolutely no idea what happened that day. No explanation at all. It was just freaky. It could have been adrenalin . . . who knows?

FREEZE IN TIME

Kate (2006) had lived in Baltimore, Maryland, for most of her childhood and used to take the bus to her elementary school.

One day me and my friend saw something that could only be described as a freeze in time. Just as the bus was turning around the corner, we saw a man standing next to his car, frozen in time. And then, light rays coming through the trees kind of crystallized.

No one else saw it, maybe because they weren't even looking, but we both saw it again and again thereafter, just as the bus would turn that corner.

Another strange thing is that the bus seemed to slow down, then speed up as we went by that corner, like some cool movie effect.

DISTORTED TIME

Nicola, too, wrote about her experiences with time distortion.

This strange thing happened to me on March 15, 2004.

I finished work at 4:00 P.M. and was waiting at the top of a side street to cross both lanes of traffic to get onto the main road.

There's always quite a long wait there, and there was again that day. I was looking left at the oncoming traffic when all the cars that were traveling at the usual speed suddenly slowed simultaneously.

It wasn't like they were slowing to stop for something either. It was like time had slowed or stretched.

So I glanced right and saw the traffic was traveling at the usual speed. I looked left again and the cars were still moving in slow motion.

Then all of a sudden the cars surged forward as if breaking free and everything was normal again. I sat there for a little while, missing a couple of opportunities to pull onto the main road, wondering what I'd just witnessed.

AREA 51

John Holmgren (2008) wrote about his TWIDDER eleven years after it happened.

> In 1997, my wife, two boys, and I were living in the small town of Rachel, Nevada. Rachel is located 10 miles north of Area 51, a top secret military base where the U. S. government conducts all sorts of experiments. Area 51 builds top secret aircraft, like the F-117 Stealth, the SR-72 and 75 aircraft, and unusual aircraft with lifting bodies and no wings. Sector 4 is located six miles east of Groom Lake. Groom Lake is part of the whole Area 51 sector.
>
> I ran a small t-shirt shop in Rachel named 'Close Encounters T-Shirts.' The town's population is 98, so the town is fairly small.
>
> One morning in March of 1997, I opened the t-shirt shop as usual. It was a clear day and the temperature was around 65 degrees. My wife would occasionally come into the shop and help me with the customers.
>
> Around 11:47 A.M., she came in and asked if we could have lunch together. I replied to her that I would, but let me lock up the store first. I shut everything down and locked the doors.
>
> We both got into our 1983 Honda Accord and drove to the gas station in Rachel. It was only a block or two away. We

grabbed a hamburger from the floor freezer unit and heated it up in the microwave. After it was warmed up, we stayed about 20 minutes and ate lunch there in the gas station.

The time was around 12:15 P.M. when we left the gas station. I had decided to cruise around the town because I was bored and didn't want to return to work until around 12:30 P.M.

The town consisted of a few homes and a few mobile homes and that's about it. We drove down one of the back roads and stopped at a stop sign.

For some reason—and I don't know why—I paused at the stop sign longer than I should have. The air felt strange around us, almost thick and heavy. I waited for about two minutes and proceeded back to the t-shirt shop.

I parked the Honda and my wife and I got out. I stood by the side of the car and sensed something was wrong but couldn't put my finger on it.

Within a minute or two it hit me, and I realized what it was. I asked my wife if she noticed something strange. She looked around and then gave me the strangest look.

Her reply was this, 'Honey, the sun has gone down. What the h---is going on here?'

I looked at my watch, and the time read 7:30 P.M.! At that moment, reality for me changed. I lost seven hours of time. My boss, Don, around the time I was looking at my watch pulled up in his truck. My boy and his boy were inside my motor home playing video games. I tapped on the motor home window and my son came to the door. I asked him what time he thought it was.

He said around 1:00 P.M. I said it was 7:30. He said I was nuts until he went outside and saw the sun gone. I told my boss what had happened, and he began to laugh.

My reply was this, 'I knew you would think I lost my mind.'

He said, 'Wait until you lose three days, like I did!'

Upon further investigation, I found out that in Rachel, Nevada, missing time is normal. Rachel is located on Highway 375. If you want to experience a time shift, you just may there, if the time is right (p. 24).

VANISHED TIME

Maggie H. wrote about some missing minutes on the October 2008 Paranormal Phenomenon website.

Several years ago, my daughter and I were getting ready to go to work. Every other day we were snappy at each other because both of us were almost late—there was a lot of screaming involved to get my teenage daughter ready in the morning.

For the first time in our lives we were ready early. Before that day, I watched the clock obsessively—I looked at the clock on the microwave, the coffee maker, the living room clock, the bedroom clock, and I kept time by the morning shows I watched. When *Get Smart* ended, I had to turn off the television and leave for work.

I was watching TV in the den waiting for my daughter when she cane in early. We sat across from each other talking and laughing until *Get Smart* went off. We were both ready early that day *and* in a good mood (which never happened because we weren't morning people).

As soon as we turned off the television, we picked up our things and walked 25 feet to the car. We drove on a gravel road that we normally avoided. It was a street that was rarely traveled. We chatted happily all the way to work where she would drop me off and she would drive the car to her school.

When we reached my school, it was quiet with no traffic around. Kids and teachers should have been outside, but no one was around. Parents should have been dropping kids off, but there were no parents in sight.

I was shocked and asked my daughter how we got there too early. She turned to face me and said that we weren't early but were very late.

'Look at the clock,' she said calmly.

The clock showed that an hour had passed. The trip from our house to school took 15 minutes, but that morning the trip had taken an hour! I wonder about that day, but my daughter has never wanted to talk about that morning. She chooses to act as if nothing strange happened that day (p. 2).

THE LOST AFTERNOON

An anonymous contributor to the Paranormal Phenomenon website shared this TWIDDER. A young medical student was spending a weekend at the beach with several of his colleagues. They arrived together at the hotel, dropped off their luggage and went off on foot for some beers.

He recalls that he was walking behind the others, talking to one of his friends around 2 P.M. Quite suddenly he found himself at the beach, and it was very dark. His first reaction was confusion, as he knew he was lost, followed by anger, sure that his fellow students were playing a joke on him.

When he finally made it to town, he hired a cab. Because he couldn't recall the name of the hotel where they'd dropped off their luggage, it took him two hours to find it.

When he finally entered their hotel room, he was very upset and told his buddies that their joke was not funny. As he demanded that they reimburse him his cab fare, he noticed the puzzled expressions on everyone's faces.

One of them said, "Where the h--- have you been? We've been to the police, to every hospital, and we've looked for you for ages!" They were every bit as upset as he.

Apparently, what for him had been a blink of any eye, for the others was actually five hours. To this day, none of them can explain how he "disappeared" while talking to his friends and appeared at a different place . . . and where he actually was during those five hours.

THE BATHROOM BREAK

Terry D. Burgoyne wrote of a lost time episode while he was an art school student.

I have had only one experience of what I always refer to as a possible time-slip though I "missed" only thirty minutes. This was back in about 1952/3.

About five minutes after an important class had begun, I had to apologize to the teacher and to the many other students in the room as I just HAD to go to the toilets, which were situated at the far end of a long, narrow locker room. (It was really not a room at all, but a corridor leading to the washroom.)

The door of the classroom was only about five paces from the entrance to this corridor over which hung a large 'post office' style clock in its wooden casing and brass trimmings and on which the hours were marked in Roman numerals.

I always loved this clock, as I do all old time pieces, and always looked at it to check the time every time I passed it. On this occasion, I saw it was twelve noon.

I went straight to the washrooms and spent at most three minutes there and then hurried back to class again.

On the way back along the locker room, I looked up at the clock again out of habit and saw that it was just past 12:30 P.M. I thought that was curious as I hadn't dawdled in the washrooms.

When I got back into class, the teacher looked at me very sternly indeed, and several of the students demanded to know where I had 'been all that time.'

I said I just went to toilet. Sorry! My schoolmates took this as some kind of joke and made remarks about 'how long some people take to do some things.'

The teacher called them to order and we finished the class . . . the final 25 minutes of it, at least. I have never been able to explain that loss of half an hour.

Chapter 7.

Instant Replay

Instant replay is a technology that allows broadcast of a previously occurring event.

In the 1999 science fiction comedy film *Galaxy Quest*, the good guys got a second chance to save the crew through use of the mysterious Omega 13, a device that allowed the user to travel back exactly 13 seconds into the past, relive the present, and change the future. (Would that I'd had access to the Omega 13 the year I came within two numbers of the winning lottery ticket. Instead of $140,000, we went home with $20. Oh well, that bought us dinner at McDonald's and an evening at the Dollar Movie Theatre. Better than a sharp stick in the eye any day.)

TWIDDERS experiences centering on Instant Replay are eerily similar to the effects of the Omega 13 or a sports rewind of the winning football goal. But unlike the sports replay, which can be shown ad nauseum, a TWIDDERS replay is a same-song-second-verse-one-time-only event.

One more comment: if So Slow is a phenomenon caused by a wrinkle in time, then is Instant Replay a fold in time?

A MILE BACK

Glenn (2002) shared his reality blip from 1991, when he was a university student in Nova Scotia. What began as an ordinary bus trip back to his hometown to visit his parents turned into a bizarre episode of Instant Replay.

I sat at the back of the bus and there was nobody around me, but there was a family sitting behind the driver in the front. The bus ride was uneventful until we came close to my parents' hometown.

I was looking out the window and looked at the Michelin Tire Factory as we went by it going uphill. When the bus reached the top of the hill, I got a strange feeling, and for some unknown reason I started to imagine many people on the bus laughing at me!

Right then there was a blip in reality, and the bus was suddenly about a mile back on the highway! I then had the experience of watching the bus drive by the tire factory again!

This kind of scared me, and I noticed that the family sitting in the front who were talking loudly before were now dead quiet.

I approached the bus driver when we stopped and told him what I thought happened. He looked really nervous and he said, 'Things like that happen.'

MR. HAWKINS

We're used to time progressing in a linear fashion, one event leading to another. Strangely, it doesn't always work like that. Consider the TWIDDER of Eula White (2004) that occurred in rural Alabama in the 1920's.

She told a lot of stories of the people and events of those days; most of them were of interesting but ordinary events. But one day she told her daughter a story of an unusual event that she had directly experienced as a young girl along with about a dozen other women and children. "I remember this event well even after all these years," she said, "precisely because it was so unusual."

In those days rural Alabama was still kind of backward. There was little electricity, and horses and wagons were the only transportation for many farm folk.

I remember it was a bright summer day. Early that morning the other women and I had gathered on the front porch of the Hawkins' farmhouse to shell quite a few bushels of peas and beans.

About mid-afternoon we were still on the porch shelling peas. We looked up and saw Mr. Hawkins approaching the house. Thrown across the saddle in front of him was a large white cloth sack of flour and cradled in his left arm was a brown bag of other groceries.

We watched as he rode up to the gate, and he stopped there, waiting for someone to open it. One of the boys ran to the gate and opened it. Then in full view of all of us women and children, Mr. Hawkins vanished! He just disappeared, instantly!

We sat there for a second or so, just astonished. Then, terrified, we began screaming. After a few minutes, we calmed down but were still shaking and confused. We just didn't know what to do. So after a while we went back to shelling peas. Mrs. Hawkins made one of the boys go close the gate.

About a half hour later, we looked up and again saw Mr. Hawkins riding toward the house with that same white sack of flour across the saddle in front of him and that same brown bag of groceries in his left hand.

Again he rode up to the gate without a sound and stopped. None of us had the nerve to open the gate. We were all just too afraid to move.

We just sat there staring at him, waiting to see what would happen next. Finally, to our relief, Mr. Hawkins spoke, 'Well, is someone going to open the gate for me?'

Mr. Hawkins got there before he arrived (p. 15)!

THE CAR

Ryan Bratton shared a similar tale September 2004 of an Instant Replay that he witnessed at the tender age of eight. It was an otherwise ordinary day for him and his friend as they were sitting in his yard while other kids rode their bikes up and down the driveway.

A car came down the road and stopped at a house. A kid got out and ran inside making noises like kids around his age always make. Then a girl rode her bike down the driveway.

A couple of minutes after this happened, the SAME car went down the road, stopped at the house, and the SAME kid got out of the car and ran inside screaming the EXACT things he'd been saying. Then the girl went down the hill on her bike AGAIN.

I looked over to my friend, and he said he had no idea what had just happened (p. 26).

SAME DAY TWICE

Peruse your memory banks for the plot of the 1993 comedy Groundhog Day. In the film, Bill Murray plays Phil Connors, a Pittsburgh TV weatherman who finds himself repeating the same day over and over again.

James G. (2003) had a replay of a 24 hour period in the summer of 1997. Unlike the hapless Murray, doomed to repeat the same day until he'd worked through his life priorities, James just had the incredible experience of a singular Instant Replay.

I actually lived through the same day twice. I awoke one Wednesday morning, went to work, and came home with no incident. There was nothing unusual or profound about the day.

The *next* morning I awoke and began my pre-work routine which included having the TV on the 'Today Show.' They gave the date at the beginning of the broadcast, but I wasn't paying attention at that point.

When I got a cup of coffee and sat down to watch, I realized I was seeing the same headlines from the day before. These were not follow-up stories but the same exact story word-for-word. This seemed a bit weird, but I figured there was a problem at the national or local station and they just accidentally ran yesterday's show.

Well, the show continued well past any time that a mistake would be detected. I went to work, and on the way I heard the same jokes and interviews on a morning show I listen to.

At this point I had feelings of anxiety because this was indeed strange. But I actually have always wanted to be able to relive points in my life to make improvements. As I mentioned, nothing profound happened that day.

On the way home I stopped at a gas station and bought a paper and it was *yesterday's news*. This has never happened since, and I keep wondering how and why it happened (p. 19).

THE BOARDWALK

Erin T. (2005) wrote about a freezing cold, dark December day back in 1996. She was 16 years old, and on Christmas vacation in New Jersey with her dad and step-mom. They were visiting family in Ocean City, New Jersey.

> My dad and I decided to take a ride over to the boardwalk and explore the barren shops and carnival rides. My family and I always loved to sneak into places that seemed like you shouldn't be there.

> We bundled up like bears and made it to the boardwalk just as it started to flurry, so we wrapped our scarves and pulled our coats a bit tighter.

> There was not one other soul on that boardwalk, no one for miles; you could feel the emptiness of the place, but we loved it. We were just walking and talking, dad and daughter . . . all was good.

> I have to admit, if I didn't feel the safety of my dad, I would have been pretty spooked out. The atmosphere was pretty creepy; dreary, snowing, old carnival rides, nothing but the sound of wind whistling past your ears, and the bite of the cold on your fingers.

> We had walked about a half mile up the boardwalk when we noticed someone coming toward us. We stiffened up a bit only because we had been totally alone, messing with closed rides and thought maybe someone was coming to kick us out or yell at us or something.

But the closer he came, we realized it was just someone jogging. We watched him come toward us. As we were getting ready to say hello, he passed and we smiled, and he just kind of stared at us.

And then something totally bizarre happened. About four seconds later, I looked up and that same jogger was about 20 feet ahead of us again, coming toward us, like he had never passed.

I was very confused at first, so I looked behind me to see if the first jogger was still running, but there was no one behind us. I could tell my dad was going through the same questioning because he was looking behind us too. So the jogger coming toward us was passing us again, the exact same guy, with the same clothes, the same strange empty stare, the same guy.

At first dad and I didn't say anything. I think we were both trying to make sense of it in our heads logically. But I couldn't. There was no way he could have run back, passed us, turned around and come back toward us in such a short time, without us seeing.

We were the only people for miles, and he was the only other person we'd seen, at all, period. If it was a second jogger who looked exactly the same, then we would have seen him coming along with the other jogger; he was only about 20 feet behind the first jogger. We also would have seen the first jogger behind us when we turned around.

After that, my dad and I got a bit creeped out, so we quietly walked back to our car, on the street instead of the boardwalk. To this day we both think something supernatural occurred; we just don't know what it was (p. 12).

THE DICTATION

On June 2009 'Hensnchicks' wrote about an Instant Replay that happened to her in Rochester, New York, in October 2002. She had been doing medical transcribing of dictation in her home for several medical providers in the surgical practice where she also worked as a nurse.

> Late one night when the taped dictation had kept me exhausted at my computer for many hours, I finally reached the last letter on the last patient. On the tape, the doctor described how 'Mrs. Smith' was found to have a certain condition and that she had opted to have surgery to correct it.

> The next day, I took the transcribed dictation to the office, and it was placed in the patient's chart. That next week I assisted the same doctor who had dictated the letter.

> During the shift, we saw Mrs. Smith. The doctor examined her, and they discussed her condition. Afterward, the doctor pointed out to me that somehow an office note had found its way into Mrs. Smith's chart . . . it detailed her condition and that she had opted for surgery, but the note was written a week before the patient had been seen!

> [In fact], Mrs. Smith hadn't been seen for three years [prior to this day's visit]! The description in the chart note was accurate, except it said she opted for a surgical solution. In reality (or should I say *this* reality), the patient had opted for conservative treatment and not the surgical option.

Why was the letter dictated to me on the tape detailing a patient who hadn't yet been seen? And how could it have been correct in her diagnosis when the patient hadn't yet been seen?

FINAL APROACH

In 1994 Martin Caidin wrote *Ghosts of the Air: True Stories of Aerial Hauntings*, a compilation of out-of-the-ordinary flying accounts. An aviation legend, Caidin is also a space technology and science fiction author. In fact, his best-selling book, *Cyborg*, spawned the 70's television show "The Six Million Dollar Man."

One of the strangest occurrences recounted in this book involved a passenger jet and an airport. Caidin omitted specific dates and names of this TWIDDER by request of those involved. However, it happened sometime between 1964 and 1980 when National had the 727 in its fleet.

The Boeing 727 of National Airlines (a company now relegated to history) was in the long slot for landing at Miami International Airport in Florida.

Everything aboard the flight was normal, and the aircraft was functioning perfectly. The pilots followed their orders from air traffic control, turning when the calls came and descending on assigned corridors. Miami radar had the 727 and other targets on its scopes.

Suddenly the blip on the flat scope that represented the National flight vanished from its electronic position on the scope. One instant it was there; the next it was gone.

Several things could have caused the sudden blip disappearance. It could have happened due to an electrical failure of the transponder system of the Boeing, an unexpected glitch in the whole radar system, or even a crew member turning off the transponder that received and boosted back the radar signal from Miami.

Or it could also mean the 727 was down; that it had plunged from its approach path, for reasons yet unknown, to slam into the swampy ground far to the west of Miami International.

Immediately the alarms sounded. The reactions were automatic. Word went out from Miami Approach and Departure, and the tower, for all aircraft in that sector to 'look for a Seven Two Seven that's gone off the scope.'

Nothing. She was *gone*.

Miami Approach hit the alarm signals to the Coast Guard and other rescue forces. Choppers bolted from the ground and raced to the last-known position of the National 727.

Nothing.

Then precisely ten minutes after the radar blip vanished from the set scopes in Miami Approach, the blip reappeared before the astonished eyes of the people now crowded around that radar position!

Reappearance was strange enough. But this was ten minutes after the 727 had disappeared, and now it had reappeared in exactly the same position it held in flight when it vanished, both on radar and in flight.

The 727 pilot continued talking with Miami Approach, and then Miami Tower, in an absolutely calm voice. Nothing unusual could be discerned from his tone or his words.

The still-astounded radar operator worked the 727 in closer and then handed off the airliner to the tower for final landing instructions. The 727 slid in, flaps extended and gear down, and made an absolutely normal landing.

The airliner was directed to park in an area separate from the terminal gate. When it stopped and the doors opened, federal investigators and officials of National couldn't get into that jetliner fast enough. The crew regarded with some astonishment this unexpected and unexplained flurry of activity and barrage of questions.

Then they were told what had happened. "You disappeared from the scope on your descent. For ten minutes there was no radar picture of you people. When you came back on the scope, your position was exactly where it had been. Not only that, but several airliners *flew through* the space you had occupied in those ten minutes. What happened up there?"

The crew and the passengers who were questioned were in a different situation. Something incredible had happened to them, and they didn't know the first thing about it.

"Nothing happened," the captain insisted. "Nothing out of the ordinary, that is. We were on approach, we came in, and we got the tower, and we landed. Period!"

"No break in communications?"

"None."

That only opened the door to another puzzle that baffled everyone involved. A member of the flight crew looked at his watch. He checked his watch against those of the rest of the crew and then against the clocks of the aircraft. They all matched, but every watch and clock on board that Boeing 727 was ten minutes *behind* the watches on the ground.

Where had that airplane gone for ten minutes? Or was this an instance of Instant Replay (pp. 223-226)?

BILOCATION?

A wife known by her I-Net moniker of "MOMDBOM9" (2009) wrote of an episode of Instant Replay that occurred one very hot July day in 2005.

> I was sitting in the family room of my one-story brick ranch located in O'Fallon, Illinois. It was 11:45 A.M., and I had just finished watching a video and was contemplating eating lunch at a nearby Chinese restaurant.

My husband, Mike, had been at his church for a workshop since early in the morning. He had told me there would be a luncheon served after the meeting.

I looked up at the clock and noted the time. I then saw my husband walk up our sidewalk (from the driveway outside of the family room's window) to adjust the outside faucet in order to water the grass and ivy.

I watched him as he walked through the ivy, twist at the faucet, then shake his hands off [from] the water that had gotten on him from turning the faucet.

He stood a moment watching to make sure the water was trickling across the ivy. He wore a yellow short-sleeved shirt, khaki Dockers, dark suspenders and his casual work shoes.

At this time as I was watching him, my two Boston terriers raced across from the family room into our living room to the front door. They began barking excitedly in the entryway hallway. I assumed they heard him and knew he would be shortly approaching the door. I watched him a good three to four minutes while he fiddled with the faucet.

Absentmindedly I thought, 'Darn, he doesn't like Chinese food, so I will have to cancel my plans for lunch.'

Nothing seemed out of the ordinary to me at that moment, [but then suddenly] I literally started shaking.

Fear ripped through me when I realized I had not heard my husband's car. (He always locks it with a device on his key chain that beeps indicating the car is locked.)

I had not heard the faucet turn and squeak, nor did I hear any sounds of water running, which can be heard every time my husband activates it. (The window from which I had observed my husband is full-length and about four feet from where I had been sitting.) I realized I could not have missed the distinctive sound of the faucet and water.

Additionally, I began seriously shaking when I realized that the outfit my husband had been wearing while watering was not the outfit he had left in that morning. He had been wearing blue-jean shorts with a short-sleeved, blue-checked shirt and athletic shoes with white socks. Very anxious with this thought, I turned to see why my dogs were still making such a racket and were still barking at the front door.

When I turned my head back to the window after a brief time, Mike was gone. I had a clear view of the front of my yard, the driveway, sidewalk and street. He just vanished.

The dogs became quiet. Totally afraid of what I had just witnessed, I started trying to piece everything together in a logical manner. Mike came home, changed his clothes, etc. . . . My mind was churning over what I had just witnessed. How had this been possible?

I immediately grabbed the telephone and called my husband on his cell. He answered right away and told me he had just sat down for the luncheon; he had never left the church.

I asked him what clothes he had on, etc. He told me shorts, etc. I related what I had just witnessed. My husband of 32 years took this very calmly.

He really didn't react because he had no explanation for the event. (My husband is a civil engineer with typical, logical left-brain thinking. Odd occurrences do not fit in the realm of science as he knows it.)

I consider myself an educated, mature, ordinary person, but I have no reason for this to have happened. I still get shivers when I think about this today.

THE MONITOR

Like the former woman, wife and mother, Sheri N. (2009) wrote about a TWIDDER that shook her world.

As usual, the long work day was coming to an end, and I was dutifully putting the last load of laundered clothes away in our bedroom when I heard a ruckus on the baby monitor just a few feet away from me.

I thought it strange when I knew my husband and toddler were both in the living room quietly watching TV as my two year old drifted silently off to sleep curled in my husband's lap as he caught the evening news. The bedroom door was straight in front of me and I could see all the way down the hall to my husband and son in the Lazyboy chair as this ruckus over the monitor continued.

It didn't take long for me to realize the sounds were very familiar. Earlier in the day, I was in my toddler's bedroom putting a load of folded clothes into the drawers and picked up some stray toys and books that weren't being played with at the time. As I was doing so, I was telling my son about the story of 'Jack and the Beanstalk' for the first time.

Now I stood in disbelief as I heard the drawers being pulled open and shut and rustling of the toys and books being put into their proper places. But I nearly fainted when I heard my son's voice over the monitor!

I kept looking back and forth at my husband and now-sleeping son in the chair in the living room, and the monitor sitting on my dresser that was literally replaying the specific events from earlier in the day!

The monitor is a standard baby monitor bought from Walmart and is NOT a recorder, but instead monitors the sounds coming from the room as they are happening at the present time only.

I listened as my voice retold the story of 'Jack and the Beanstalk' and listened with familiarity as my son responded in baby-talk to the tale he had never heard before. The incredible part was this all happened five hours earlier on the same day!

I quickly called my husband into the room as he listened to the last part of the story with my voice coming through the monitor and our son's coos and chuckles. He stood stunned and turned his head and looked at our sleeping son flopped peacefully over his shoulder.

In disbelief, he asked, 'How in the h--- . . . ?!,' as his voice drifted off trying not to miss a thing. I just stared at him in the same disbelief and we both just shook our heads.

This has never happened before or since and became pretty clear from the beginning that we were listening to some kind of warp in time. I never imagined in a million years that I would be witness to it and must admit, if it should happen to you, it is indeed one of the most incredible moments one can ever experience!

Chapter 8.

Lost Locations

"Yes," he said. "Witches have known of the other worlds for thousands of years. You can see them sometimes in the Northern Lights. They aren't part of this universe at all; even the furthest stars are part of this universe, but the lights show us a different universe entirely. Not further away but interpenetrating with this one. Here on this deck millions of other universes exist, unaware of one another . . ."

He raised his wings and spread them wide before folding them again.

"There," he said, "I have just brushed ten million other worlds, and they knew nothing of it. We are as close as a heartbeat, . . ."

—Kaisa, *The Golden Compass*

Wormholes (tubular shortcuts in space-time) may transport you from one spot in space to another. Both ends of a wormhole could be intra-universe (existing in the same universe) or inter-universe (existing in different universes, serving as a connecting passage between the two.)

And, as a select few of the following accounts record, inanimate objects may be as likely as humans to travel through a wormhole to become part of a Lost Location TWIDDER.

Hmmm.

In light of modern physics positing multiple universes, as well as a means of traveling between them, I have to tell you what really puts me into a cold sweat in the middle of the night is the thought that a Lost Location experiencer might get stuck elsewhere.

What if my young family had decided to camp overnight on that rocky beach on the Oregon coast? Would the westerly road that delivered us to the beach and that we were never able to find again, have disappeared as we slept, imprisoning us forever in a world not our own?

Or will the universe always boot us back to the plane of our birth—even should we try to stay?

THE MEADOW THAT WASN'T

In January of 1998, Richard Alvarez recorded a Lost Location TWIDDER from the early 1960's that still sends chills up my spine.

In a remote mountain camp that I used to frequent, a trail led down the river a short distance, I think about a mile or so, to a Rotary Gospel camp.

Both our camp and the Rotary Gospel camp and the river between them were in a narrow canyon with high ridges on both sides and no side-canyons. Several of us had been down that canyon, also up and over the ridges on both sides, and we knew that there was no way to leave that canyon. Although the terrain was steep, hiking was easy along the trail, and that route was considered quite safe for youth day-hikes.

Generally, it is hard to become lost when following a river downstream. But when following a river upstream, in some cases there is a possibility of becoming lost by accidentally following a tributary rather than the main stream.

In this case though, in that deep narrow canyon, all that they had to do was hike back up the canyon to our camp. And in an emergency, they could have gone into the Rotary Gospel camp and used their telephone. So we considered it a safe hike.

One afternoon two adults from our camp, with several youth members ranging in age from maybe eleven years to about fifteen years, hiked down the trail toward the Rotary Gospel camp, intending to turn back at the edge of that camp and return to our camp well before dinner.

The adults and some of the youth members knew the trail well. In that narrow canyon, over the short distance to the Rotary Gospel camp, and with several people who knew the trail, a map and compass seemed unnecessary.

Several people had wrist watches (the mechanical wind-up kind; modern quartz crystal watches weren't available until 1969). One person had a pen-light, and one youth member had a camera.

Some distance down the trail, when they expected soon to reach the Rotary Gospel camp, the canyon widened into a huge meadow that nobody recognized. Nearby at one edge of the meadow beside the trees there was a log cabin, but no people were visible.

The meadow seemed much too wide to fit in that narrow canyon. There was no possibility that the group had passed the Rotary Gospel camp as it was in the bottom of the canyon.

Although they knew that something was wrong, they followed the river into the meadow, alert for any tributary streams that might confuse them when they retraced their route. After a while, they noticed that the river had decreased greatly in size and flow.

That alarmed them, and they turned around and headed back up the river. Even as they followed the river upstream in the meadow, the river rapidly disappeared.

There were no dams on that river, and no recent weather patterns to account for such a sudden decrease in the flow. They followed the sandy river channel back upstream through the meadow, but soon even the channel disappeared, leaving the group in that big flat meadow with only the mountain ridges for land-marks, and even those ridges seemed to be much too far apart. From the group's position within the meadow, they could not even see the canyon where the river had entered the meadow.

Since they did not have a compass, one of the adults used the sun as an indicator of azimuth to keep them headed back toward camp. But as they took frequent note of the sun's position, they noticed that it was setting much too fast even though their watches indicated that the time still was early afternoon. At that time of the year, the middle of June, at that location, the sun sets very late.

Fortunately, they had not gone far into the meadow before they turned back. They reached the edge of the meadow just about when the sun set, and there they suddenly came upon the canyon that they had hiked down.

It had been hidden from their view (or maybe it was not even there?) while they were in the meadow. The river was flowing normally in the canyon, and it sank into the ground where it entered the meadow. When the group started up the river, they immediately recognized familiar land-marks, to their enormous relief.

In that deep and heavily forested canyon, soon it was so dark that they barely could see where they were going even though it still was early afternoon according to their watches. With their one little pen-light, they stumbled over the rocks, expecting to find the smooth trail that ran beside the river.

But apparently they missed the trail in the darkness. Rather than spend the night there, they decided to continue the short distance up the river to our camp.

As they explained later, after having the canyon and the river disappear right out from under them and becoming lost in the meadow and having night come in the early afternoon, they were so shaken that they wanted to return to civilization right then.

They no longer even were sure that our camp still was there or even that morning would come, and they did not want to spend eternity in a canyon that maybe went from a non-existent meadow to nowhere. So they continued stumbling over the rocks even after the little pen-light's batteries were exhausted.

Hiking on a smooth trail is easy in the day time. Even hiking over rocks in the river is not too bad when you can see where you are going. But it took them hours to go probably less than a mile up the river to our camp.

Some of them were badly bruised from falling over the rocks, but fortunately nobody was injured seriously. They were enormously relieved to reach our camp buildings and our camp's own little meadow. They headed straight for our group's campsite.

179

I had been out of the camp that afternoon; otherwise, probably I would have gone with them on that hike. When I returned to the camp that evening, the decision already had been made not to start searching for them until morning.

The camp director already had telephoned to the Rotary Gospel camp and had determined that our group had not been seen there. When they returned, I was almost as relieved as they were.

We all stayed up late into the night with the camp director, going over their story. None of us found any significant discrepancies between their accounts. Their watches all indicated late afternoon.

At first, they were not hungry; to them, it was not yet dinner time. Apparently, at least in that meadow, time had affected their environment at one rate and had affected both them and their watches much more slowly. Or maybe they still were too shaken to be hungry. Part way through the discussion, the camp director served them dinner, and we finished the discussion as they ate.

The next day we had everybody who had been on the hike write their accounts on paper with no further prompting and with no discussion between them as they wrote. After they finished writing, we asked them to add their recollections of various details, again with no discussion as they wrote.

I remember asking them to record details of the transition between the canyon and the meadow, and of the river bed in the meadow, also about where the sun set over the ridges above the meadow.

I do not remember now what details the other adults requested. That process went right through lunch, and they ate with one hand while writing with the other hand. No, we were not slave-drivers; those people seemed as anxious to record that incident as we were to have it recorded.

Then we turned the youth members loose along with the two adults who had been on the hike, and the camp director and I stayed late into the evening comparing the written accounts. There, too, we found no significant discrepancies.

To the great regret of all of us, the people had been so shaken that the person with the camera had not thought to take pictures of the meadow. Probably he still is kicking himself for that.

Or maybe it is just as well that he did not take pictures; just possibly, recording that scene might somehow have trapped them there permanently. Of course, that is pure speculation, wild speculation at that.

The following morning the camp director, the two adults, who had been on the hike, and I started down the river for another look. This time we took maps and compasses, watches, cameras, flashlights with spare batteries and bulbs, and note paper.

I took a roll of surveyor's flagging (brightly colored plastic tape) for temporarily marking our route, both so that we could find our way back in case of difficulty, and, to be truthful, so that just in case we did not return, other people could follow our route. (That last thought scared all of us a little bit.) On general principles, we took lunches. Portable two-way radios were not common in those days;

otherwise we would have wanted them.

I took a small portable instrument for measuring the sun's elevation angle. For any location and time of day, it is easy to compute the sun's azimuth and elevation angle. Also, if you measure the sun's elevation angle at two known and widely separated times during a day, then in principle you can compute your location (latitude and longitude); that is a standard procedure in celestial navigation.

If we came to the meadow again, then I would compare the sun's azimuth with a magnetic compass and measure the sun's elevation angle at frequent intervals according to our watches. Then back home (assuming that we survived the trip), I would try to make sense out of any strange behavior such as the group had observed two days previously.

I also took a telephone lineman's test set; that is a little self-contained telephone like you see on linemen's tool-belts, except that test sets had miniature rotary dials in those days rather than modern push buttons.

Our camp's telephone line ran beside the river to the Rotary Gospel camp where the telephone cable ended. Along the river between the two camps, our telephone line was attached to trees at about head height, so we would be able to use it in case of emergency.

I admit to some concern that if we did find the meadow, then the telephone line might neither cross the meadow nor go around it, and thus we might be isolated. In our camp I had asked the hikers whether they had noticed whether the telephone line crossed the meadow, but none of them had noticed the line even in the canyon. As I look back on it now, we were asking for trouble. Maybe we should have had more back-up before we started investigating.

Before we left our camp, I found an excuse to telephone to one of my friends at the Rotary Gospel camp, and I arranged to meet him there. From our camp director's call two days previously, my friend already knew that something was wrong. I begged off explaining until we met personally.

Now after all that build-up, here comes the embarrassing part of my story. We walked down the trail to the Rotary Gospel camp and saw absolutely nothing unusual. On the way, I tied strips of surveyor's flagging to tree branches at frequent intervals.

I also kept a close watch on our telephone line beside the river; it was there right where it was supposed to be. There was no meadow. The river still was very much there in the bottom of the narrow canyon.

The sun was rising slowly in the sky just like it should be doing at that time of day. We went into the Rotary Gospel camp and met with our friends there. As my excuse for being there, I looked at some equipment that I had helped them to build.

But probably they still are wondering why we went there that day. Our camp director used their telephone to call back to our camp to let our people know that we had made it safely that far.

Then we took our leave and walked back up the trail to our camp, removing the strips of surveyor's flagging as we went. We ate our lunches in the camp dining room with the rest of the camp members. We felt relieved and also a little bit foolish. But we would rather feel foolish than be lost.

Those mountains have changed drastically during geologic time. The river systems have re-formed so extensively that it is hard to imagine how much different they used to be.

But all of that happened long before there were people on Earth. Thus even if our group somehow was transformed back in time to another geologic era (which I still find impossible to believe), that does not explain the log cabin in the meadow.

What was that meadow all about? And the disappearing river and the rapidly setting sun? Darned if I know. If that group made up their story, then they did a very good job of it, telling it and writing it with no significant discrepancies.

If they used the hike as an excuse for some mysterious detour, then they had one heck of a climb out of the canyon to get there. I think that all of us were basically honest, except, of course, for the standard campers' practice of telling outrageous lies around campfires. My guess is that the hikers were just as puzzled as I was and also were thankful to be alive after it was all over.

FROM MANILA TO MEXICO

In *Encounter Cases from Flying Saucer Review,* Charles Bowen (1980) wrote about a strange appearance. On October 25th, 1593, a Spanish soldier suddenly appeared on the Plaza Mayor (the principal square) of Mexico City.

He should have been with his army in the Philippines at Manila. Neither he nor anyone else could account for this strange appearance. He also told the people gathered that His Excellency Don Gomez Perez Dasmarinas, governor of the Philippines, WAS

DEAD!

They threw the soldier in jail as an apparent deserter. Weeks passed. Then suddenly Mexico City was full of news. The governor of the Philippines was, in fact, dead, and he died on the very day that the soldier had said weeks ago!

The most holy tribunal of the inquisition questioned the soldier, but all he could say in his defense was that he traveled from Manila to Mexico "in less time than it takes a cock to crow."

The soldier was returned to Manila, and it was found that he indeed was on duty the night of October 24th, 1593.

PATH TO . . . WHERE?

In early spring, 1984, Jim K. (2008) and his brother were busy growing up in Scranton, Pennsylvania.

There was an old park near the house we grew up in where our parents would let us ride our bicycles. This park used to be quite the attraction years ago, but at that time was mostly run down.

The gazebo roof was falling in, the zoo was closed, most everything was overgrown, and it had a highway built through it back in the 1970's. Despite this, for kids this was the ultimate in exploration.

There were still quite a few trails where we kids could do just that. One of those trails took us a short distance into the woods where it dead-ended at a small cliff overlooking a brook.

Across from the brook were more woods, some railroad tracks, and then I-81, which we could clearly see through the trees. We liked to sit there and throw rocks into the water and talk about whatever it was that we used to talk about.

This time, however, was different. We were on the path to the cliff, but for some reason we never made it. We pedaled and pedaled, finally coming to the edge of a huge drop-off overlooking a beautiful green valley.

It was a strange thing. We had explored all over this place and had never seen anything like it. Plus, where did the brook go?

Not only that, but where did the highway go? We thought about these things, but figured we had found some super secret new place to explore. We looked around for some way to climb down the drop-off, but couldn't see any easy way down.

We decided to go home and tell our father about it and get him to come back with us. We turned around and in about two minutes were back at the head of the path.

We did tell our father and he did come with us to explore. The only problem was the path never took us there again, only to the drop-off and the brook. My brother and I have talked about this on numerous occasions, and both agree to exactly what we saw. I really have no explanation for it (p. 10).

THE TREES

Two years after this experience, Brian D. (2008) wrote about trees that disappeared and reappeared. In October 2006, just east of Longview, Washington, Brian D. was hunting with his folks. Of the TWIDDER, he says, "46 deg. 12'49.61" N by 122 deg 47'8.04" W is the EXACT location where this occurred if anybody wants to check out the area.

My dad and mom were driving along a logging road that had a row of trees along it around twelve feet tall and so dense you couldn't see through them.

My dad stopped at the bottom of the road so that he could get out and go walk through the row of trees so he could see the clear cut on the other side and check for deer. He stopped, got out of the truck and was closing the door, taking a few steps while looking at the ground.

When he looked up, the entire row of trees along the road he'd been driving on were gone! No stumps, nothing. He could clearly see into the clear cut area.

This all happened within a few seconds. They stood baffled for a few minutes and my mom was scared, so they decided to leave.

My uncle lives near there and knows the spot as they've been hunting there for the last 20-30 years. My uncle drove up to check on my dad's story, being extremely skeptical.

My uncle said no, the trees were there and they were so big they were overlapping the road, hitting his truck. Much bigger than he'd seen them the previous week.

My dad is pretty grounded and wouldn't make up a story, and he even drove a friend of mine and myself nearly 50 miles just to go check up on the site a few months later.

The trees were nowhere to be found, just like when they had disappeared. According to my dad and uncle's time line, they say the trees are there, then disappear in the blink of any eye, then later in the week there again and bigger than ever, and then months later gone again. Who knows why or how this happened (p. 27)?

THE CAR THAT MOVED ON ITS OWN

Jane (2005) had a disconcerting TWIDDER that occurred while she was Christmas shopping.

All I was trying to do was get my mom a Christmas present and mail it to her out of state as the days were flying by, and it was close to 5:30 P.M. when stores would soon close.

I saw an antique store I wanted to go to but couldn't find the parking space I wanted on the back side of the street. Finally giving up, I parked on the main side and got out looking at a 'sale' sign on the sidewalk. Crunching through the snow, I walked to the end of the building and found the perfect gift.

I came out of the end of the building, turned the corner, and headed down the street toward my car—which wasn't there! At all!

My mouth dropped open and I just stared. My footprints were still in the snow where I'd gotten out, paused and looked at the sign, but no car.

It was now 5:30 P.M., and as I looked up and down the street, every store window was dark. No place was open for me to call the police and report my car had been stolen.

I began to get frantic, pacing up and down, wondering where I could find a phone. Maybe there was one around the corner! Around I walked, and no phone.

But there was a gray Monte Carlo that looked just like mine. I walked slowly up to it and it looked the same. I tried the key and the door unlocked. By now I was almost hysterical and started crying. I almost couldn't remember how to even drive back to my house.

Does anyone believe me? Nope, not many. I have trouble believing it myself! Did my 'wishing' make it so—wanting to park back there in the first place? Guess I'll never know (p. 16).

THE VICTORIAN

Howard Winters (1998) wrote about an abandoned Victorian mansion.

Back in the late fifties, I had a job with a waterproofing company in Atlanta known as Surface Coatings. I was the Florida representative, and I traveled the state from Pensacola all the way down to Miami, calling on established dealers of a waterproofing product we manufactured.

Over the years, it was occasionally necessary for me to drive up to Atlanta for some business need, and I would spend a few days with my family. Over the weekend, normally on a Friday morning, I would drive down US 19 from Atlanta to Tampa, arriving there in the late afternoon in order to be on hand for my trips that always began on Mondays to various dealers in the state.

This particular Friday morning, I was driving down 19, somewhere below Griffin, Georgia, and I was passing one little town after another, passing to the west of Macon, and on down towards the Florida line.

It was a sunny summer morning, and I slowed as I neared yet another little town where the pace of living was much slower than I was used to. As I came into the northern edge of the town, some quarter of a mile outside the city limits, I found a beautiful, deserted old Victorian mansion sitting off to the left side of the road, backed up by a still-standing carriage house. There was an ancient oak tree standing in the front yard. Its limbs reached out some 50 feet or so in all directions.

With no one waiting for me in Tampa, I decided to stop and walk around the house, and since its door was open, possibly walk into it to see how its grandeur of some seventy years before had survived.

Parking beside the house, I got out and walked around to the front yard, and after taking in the particular beauty of all old Victorians, I stepped up on the porch and walked into the foyer of the house.

The floor was covered with an old, old carpet, and there were pieces of furniture standing here and there. Over to the door to my right that opened into the parlor stood an antique chair that was used years ago to hang umbrellas and such on, with a section built into it to hold boots, galoshes, etc.

The house was cold inside, and this surprised me. It was a summer morning, and the temperature outside was in the 80-90 degree range. The foyer opened into a wide reception area, something like twenty feet wide, and from either side of the room in front of me, two matched stairs climbed in a lazy arc to the landing of the second floor above me. But it was all approaching decay. . . . Even now, 40 odd years later, the hair on my arms and back of my neck rises as I remember it.

Back outside before I drove away, I drove back to the stables and found that there were still buggies present in the stalls. I was amazed that anyone would have allowed such a wonderful old house to fall into such a state of disrepair.

I drove around to the front of the house, stopping in the bare front yard under the oak tree, and looked again at the beauty of this badly neglected, once magnificent old mansion. Then I drove on towards Tampa.

In later years, I tried again and again to find this house in this little town as I drove south towards Florida, and I could never find it again. The town was and is there, alright, but the house is simply gone.

I would have thought no more about it, except for the fact that a friend who knows of my unexpected ghostly encounters asked me about some house I had mentioned years before when I was working in Florida.

I repeated this story to her, and to my surprise, she said she had seen the same house. She too had been impressed by its decaying beauty, and she and her husband had stopped a moment to look at it.

The unique thing about all of this is that I was looking for that house after I first saw it back in 1957. She's many years younger than I am, and she's seen it in the past twenty years herself. Not just her alone but her husband as well!

I've never seen it again. I looked for it for two years. I never found it. She saw it on a trip to Florida back in 1987. Where does it go? Why is it sometimes there, and other times there's only a vacant field, and the tree, too, is missing? She states the tree was there in the front of the house in the yard just as I described it.

MISSING MANSION

Amy Ryan (2006) wrote about a similar experience.

When I was about ten, my family went to Las Vegas. I always stare out the windows on road trips, enjoying the view and daydreaming.

About two-thirds of the way there, I was looking out the window and there was a huge meadow bordered by a forest and in the meadow was [a] mansion.

The strange thing is, looking back on it, the forest was only a few feet thick going all around the house except on the road side, and then there wasn't any more. But at the time I remember thinking how beautiful it was and that whoever lived there was very lucky to live in the middle of such a large forest.

On the way back I watched out the window the entire trip, anxious, for some reason, to see the mansion again, but it wasn't there.

I have no idea why or how this happened or why I was so anxious to see the mansion again. I can only wonder if someone somewhere is wondering, *Who lives in that beautiful mansion, and why is it directly off a highway?*

THE PHANTOM STREET

Jeremy Simmons recorded this TWIDDER August 7, 1998 on the paranormal website.

In 1991, I was living in England with friends in the country just southeast of London in a small town called Oxted in the county of Surrey. One night, some two weeks after I had arrived there, I decided to take a long walk about the town to learn its streets a little better, and as no one wished to join me, I set out alone.

It was a lovely night, humid but breezy, with just enough chill to make a sweater comfortable. The moon was half full and glowing dully in the moist air, and only a few stars were visible.

I had been walking for over half an hour and was considering turning back (where there was warmth and good English beer awaiting me) when I decided to take one last detour. I walked up a short hill on a street called EastHill.

Strangely enough, the street running perpendicular to EastHill, into which EastHill terminated, was called WestHill. There is no explaining the decisions of our English cousins sometimes.

In any case, I walked up EastHill and turned right on WestHill. After walking about a quarter mile down WestHill, having seen no other roads, only driveways, leading off of this street, I decided to call it a night; the air [was] getting chilly and rather uncomfortable.

I turned and headed back for the intersection with EastHill to get back home. Instead I came to an intersection that I did not entirely recognize, but as this was a colloquial little town with its own little oddities, I thought I must have, somehow, missed it when I had strolled past earlier.

I walked down what turned out to be a dead-end street. I saw no sign for its name. Turning around, I headed back out onto WestHill and finally found EastHill, took it through the way I had come and was home not twenty minutes later.

But here is the part that made this night one I could never forget.

I remember that when I stepped foot on the dead-end street and with every step I took further down into it, I got the most appalling sensation of fear and of being out of place, as out of place as I ever could be though the feeling seemed to have no base in anything I could see. The street looked the same as many of the others on which I had traveled that night. But I soon left it, and the terror left me; and in great confusion and no small amount of apprehension, I went home.

The next day I had a friend drive me up EastHill and turn right onto Westhill, making up some story about losing my sunglasses (worthless article in that country anyway) and I looked for that street.

It was not there. I am not kidding. Later I went back and combed EastHill for a hidden entrance to the phantom street and had no luck.

I know I could not have been mistaken about turning right onto WestHill because the way to the left leads down a steep hill along a waterway and under a railroad underpass,

I certainly would have remembered that; it was not such a dark night. In any case there were no streets leading off of it either, not for two miles or more anyway.

I have never found a satisfactory explanation for walking on a street that never existed, but I can tell you that I remember nothing about it, except that in appearance it was not notable at the time.

Believe me or disbelieve me, but this happened.

THERE, NOT THERE

On May 17, 2006, A TWIDDER experiencer from McLane, Virginia, shared a Lost Location episode from her childhood. She called it "Around the House."

When I was little, my aunt was getting married. And being the only little girl in the family, I got to be the flower girl. My aunt was going to have the wedding at a church near her home in Virginia (she lived a little outside of the capitol) and have the reception at home.

Because there was so much commotion in the house, my mother kicked me out and told me to help my father with the cake. I was a typical little kid who wanted to help too much, so I ran with all my might around the right hand side of the house towards the front yard.

When I got there though, it was different. I know it sounds weird, but instead of the vans of catering services and waiters and people everywhere (that I saw through the windows earlier), there was nothing. The yard itself was different. The driveway wasn't there.

My aunt had a driveway that connected to the basement and so it is dug from the street to the house. I walked around the yard and also saw flowers that weren't there before. Confused but still very excited, I had the bright idea that if I ran around to the other side of the house, it wouldn't be there.

So I ran back the way I came and zipped to the other side of the house. And lo and behold, there was the catering van, and a whole bunch of people bringing stuff into the house.

So I went back the other way, and they had vanished (and the garage) and there were the different flowers again. I did this about ten times before I had to leave for the ceremony; and when I got back, I tried to show my mom but I couldn't.

For years, every time I visited my aunt's place I tried to do that but it never worked again.

THE HALLWAY

In June of 2006 Stefani Gilbert recorded a friend's TWIDDER that took place in San Antonio, Texas.

My friend—let's call her Sara—used to live in hotels with her mother because that was part of her mother's job. They lived in the Menger Hotel (adjacent to the Alamo) in downtown San Antonio for a while, and that is where my friend Sara had her experience.

She said her mother would work sometimes till three or four in the morning. So Sara would stay awake and explore the huge hotel. She said that one morning, really early, she was wandering around when she came upon a tall spiral staircase that didn't look like the other staircases.

There were walls around it instead of railing. She went up the staircase and was bumping along the wall, like a typical annoying little kid, trying to make a lot of noise.

Sara was about half way up when she bumped into the wall and fell in a hole. The paneling on the wall just moved right in, and she was in a pitch black hallway where you couldn't see anything.

But being the type of kid she was, she had a flashlight her mom gave her because she said most lights would be off since people were sleeping. So she took off down the hallway.

Sara walked for a few minutes and then came to a door—or another panel—because you could see light around it. She pushed it and it opened into the hotel's room where the guest pool was.

She walked in to find some staff cleaning the area, and when they saw her they ran to her, shut the door and told her to not say a word about that hallway.

They apparently knew who she was because they took her to the area where her mother was. Sara and her mom went to investigate.

When they got to where she fell in the wall, there was no moving panel to be found. Later she talked to hotel management, and they said they knew nothing about any of that. Sara even described the two staff members who escorted her to her mother, and the management said they didn't have any staff that resembled her accounts.

To this day she still doesn't know anything about the disappearing hallway. And she also found out that the pool had been closed off and the doors chained for over a year. The management had no idea how she was able to get into the pool room and get out, not to mention others doing the same.

LITTLE HOUSE IN THE WOODS

Harvest Mom (1999) (a woman who didn't want her actual name disclosed) wrote about her TWIDDER experience.

As a teen my friends and I loved exploring abandoned properties. There was a house just outside of town about a half mile off the highway that had intrigued me since childhood, and I had to go see it. One afternoon two of my friends agreed that it would be a great place to check out so we loaded up [our car] and went.

It was a typical creepy-old-farmhouse type place made of wood that had begun to rot decades ago: dirty, mostly broken windows, rickety porch. It was two stories high, plus an attic.

Getting [into the house] was no problem; the doors were gone so we just walked on in. There was evidence of other people having been there as the room we concluded to be the living room was full of empty beer cans and potato chip bags.

Uninterested in these findings we made our way into the kitchen. It was broad daylight outside, but this house was pretty dark inside, adding to the intrigue and creepiness. The kitchen was a wreck and seemed as though the mice had taken full advantage of whatever had been left there, so on to the dining room.

Finally we saw a set of stairs, the first ones we had seen so far, and up we went. On the third step from the top, I sat my soda can down, so we'd be able to remember that's where we came up . . . I'm glad I had the foresight to do that.

We meandered the top floor and found that it was much brighter and warmer up there, probably because it wasn't under the cover of the low tree branches as the downstairs had been.

We weren't feeling too afraid, just a little uneasy as we walked around until we got to a large room on the east side. My friend, Chris, left the room, stating that he was finding it difficult to breathe; [he figured it] must be the dust. My friend Megan was very pale and shaking so we decided it was time to go.

We walked around the top floor looking for stairs to the attic and never found them. We found a set going down which we had missed on the first few go-rounds, so down we went.

They led to the living room! How was that possible? We walked around the downstairs, peeking in all the little doors for stairs, only to find closets. There wasn't a door to get out of the living room, so we went back upstairs and walked around some more.

Finally Megan spotted the stairs where my soda can was, and we happily bounded down them and back into the dining room. We made out way around the downstairs again, just to be sure . . . the dining room led to the back porch and the kitchen and had stairs to go up. The kitchen only led to the dining room and the living room; the living room only led to the kitchen and the porches.

Okay, so where were the stairs we walked down? This was the same room [the living room] because the mess on the floor was the same and the mantle above the fireplace was the same . . . so what was the deal here?

We went back upstairs, and Megan stayed at the top of the stairs while Chris and I walked around again in search of the mystery steps we had gone down. We never found them.

THE DIRT ROAD

"Jim (my friend's dad) and his buddy, John," Robin (2007) wrote, "went on a hunting trip in the early '70's to Tennessee in Jim's pickup truck. After an uneventful week of hunting, they decided to return to Pennsylvania."

The road out of the hunting area was a long dirt road. It intersected with a two-lane road, which would eventually lead to the main highway. When they got to the end of the dirt road, they turned right onto the two-lane paved road.

Jim noticed a strange gray fog in his rearview and side mirrors. He said that it was as though there was NOTHING behind the truck, just grayness. Ahead of them was a normal-looking day in the mountains.

He didn't say anything to John about it for quite a while. Finally, near the turn-off to the main highway, he said to John, 'Do you notice anything strange?' John had noticed the grayness behind them, too. Neither could figure out what was going on, and for some reason, they didn't stop to investigate.

Then as they approached an intersection, they were somehow instantaneously transported back to the end of the dirt road at the hunting preserve about two hours away.

They had not driven in a circle. According to the odometer and the gas gauge, they had only traveled one way. Neither man had been drinking or taking drugs; they are both very honest, upstanding members of their community. They are in their seventies now but still remember the incident very clearly (p. 10).

THE HOLE

In December 2003 Chris K. wrote about a disappearing hole. Chris grew up in a house at the top of a street that dead ended along Mulholland Drive in Sherman Oaks, California.

I spent many hours up the dirt fire road that leads up to Mulholland itself. Approaching a tiny street that shoots off of Mulholland, there used to be a plateau that appeared to be like any other in the area, designed to hold a house.

I really liked this area because of its isolation and slightly magical feel (my father tells me as a child I used to call it the 'bat cave').

One day when I was around ten, I hiked up there exploring and found a hole in the ground. This was a VERY strange hole.

It was about 18-24 inches in diameter and appeared to be perfectly circular and vertical. The dirt in the general area was loose, but when I touched the wall of this hole/shaft, it felt hard.

About a week later, this hole was gone. I mean no trace; just gone. Like I had dreamed it. But I know I did not dream this.

I remember dropping a rock into it to find out how deep it was, and I never heard a sound of it hitting anything. Therefore, I got the impression that it was extremely deep (p. 25).

OUT OF TIME AND PLACE

J. H. Brennan in *Time Travel: A New Perspective* (September 2002) shares a TWIDDER involving a biologist and his wife. The biologist, Ivan Sanderson, spent some time with his wife in Haiti, conducting a biological survey. They were in a remote area of the island with their assistant, Fred Allsop, when their car became bogged down in the mud, and they were forced to walk home.

Allsop walked ahead. Sanderson was walking with his wife when, to his astonishment, he saw a number of three-storied houses of varying types along both sides of the road. It was night, but he saw them quite clearly in bright moonlight, even noting that they cast appropriate shadows on the ground. And so Sanderson's tale begins.

These houses hung out over the road, which suddenly appeared to be muddy with large cobblestones. The houses were of about the Elizabethan period of England, but for some reason I knew they were in Paris.

They had pent roofs with some dormer windows, gables, timbered porticoes and small windows with tiny leaded panes. Here and there there were dull reddish lights burning behind them as if from candles.

There were iron frame lanterns hanging from timbers jutting from some houses, and they were all swaying together as if in a wind, but there was not the faintest movement of the air about us.

Sanderson must have suspected he was hallucinating, for when his wife stopped so abruptly that he walked into her, he nonetheless asked her what was wrong. For a time she remained wide-eyed and speechless, then "She took my hand, and, pointing, described to me exactly what I was seeing."

She said, "How did we get to Paris 500 years ago?"

The Sandersons compared detail after detail, pointing out various aspects of the houses to one another. Then they began to feel weak and started to sway. At that point Sanderson called to Fred, whose white shirt was still discernible in the distance.

Fred ran back to them by which time the strange houses and the paved road had vanished. Fred had seen nothing unusual.

The fact that Fred noticed nothing amiss indicated only that he had gone ahead of the point where the time-slip occurred.

Following are narratives of inanimate objects apparently traversing wormholes to different locations.

THE GLASSES

On August 2008 Neil Faulks wrote about his vanishing glasses.

I am a serious-minded, 45 year old male, which is why the following event has unnerved me. I live in Birmingham, England, and on Sunday, 25th May 2008, I went up to my office at home at around 4:30 in the afternoon to continue scanning family photographs, which I have started to do lately.

I wear separate general and reading spectacles, so I took the reading pair with me. I sat at my workstation and swapped my eyeglasses over, placing my normal use pair about 18 inches to the left of me in front of my printer.

I had been scanning for around 20 minutes and decided to make myself a cup of tea downstairs. I turned to my left to get my general spectacles . . . and they were not there. I got up off my chair to look on the floor in case they had fallen, but [they were] nowhere to be seen.

I spent the next hour turning the small room inside out in increasing frustration, including turning out the set of drawers on which the printer stood but to no avail.

My wife came home just then and I told her what had happened. I sensed the look of disbelief on her face, but she said she'd look anyway. She also looked high and low, including the aforementioned set of drawers.

The spectacles could not be found. I donned a spare pair of glasses and spent the next week going about my daily business (which included several more nights in my office working and searching the room again for those damned spectacles), even going several more times into those drawers for various items.

On Sunday, 1st June, I once again went into my office to work and started at around 3 P.M. An hour and a half later, I opened a drawer underneath the printer to get something out.

I lifted a pair of bagged earphones at the front . . . and there neatly folded up were my missing eyeglasses! I really could not believe it and shouted to my wife what had happened.

We had both checked that same drawer on numerous occasions and THEY HAD NOT BEEN THERE. Even more so, I have realized that they had been missing for exactly a week (to the hour) before they turned up.

I do still go into the office even though this has spooked me a little, but I don't think I would feel the same had I seen or heard anything weird at the time (p. 6).

THE LOAF OF BREAD

Donna F. wrote in July 2009 about a frustrating kitchen experience. (If Donna lived with us, the answer to any missing food from kitchen counters would be Chewie, our Great Dane who is more than happy to make any food disappear.)

I buy in bulk to save money so I buy four to five loaves of bread at a time and freeze them. This past weekend (June, 2009) I went shopping and bought some bread, not realizing that I had two loaves still in the freezer.

So I moved those two loaves to the front of the freezer to use first. I finished a loaf yesterday and pulled out one of the two loaves to defrost. I always put the bread on the counter next to the refrigerator to defrost.

My family left the house and locked everything up to go to work and summer camp. At night when we got home, I had to hurry and change clothes to go to a work-related dinner. I came home and can't remember if I saw the bread sitting there or not.

This morning, though, it was totally gone. I was confused, knowing that I had left it there. Just to make sure that my mind wasn't playing tricks on me, I looked in the freezer, and sure enough, there was only one loaf of bread in the front of the freezer.

I looked all over the kitchen and living room (thinking that maybe my husband moved it to eat with dinner the night before). I went upstairs to my seven year old daughter's room, thinking that maybe she got hungry in the early morning and grabbed the bread (even though she's never done that before, and we have Pop Tarts for her to eat).

I asked my husband, and he said he didn't do anything to it. I asked my daughter, and she didn't know what happened to it.

I looked in every cabinet and [the] stove to see if I might have, without thinking about it, put it somewhere weird. The bread is gone. Our house really isn't that big, so a loaf of bread would stick out if it was somewhere it wasn't supposed to be.

I'm very confused. I can't imagine that someone would break into our house and steal only a loaf of bread. I guess it's one of those things that can't be explained.

Chapter 9.

Alternate Reality

I've spent the last decade teaching high schoolers. Such a job would make mush of the staunchest brain, and mine was a long way from firm to begin with.

Nevertheless. I had an experience a few years ago that still bugs me, and I don't think it was a case of early onset dementia. . . .

It was the spring of 2008, and I couldn't get her out of my mind. Not a day went by that I didn't think of this young gal who I couldn't-for-the-life-of-me remember how I knew. I could see her face clear as a bell: heart shaped with large, deep set brown eyes, softly curled short brown hair, generous lips, a gentle disposition, and a great worker. I could even recall the tenor of her voice, but I could not remember her name or how the heck I knew her.

Yet I kept wondering how she was doing. Surely she'd been a student at the high school where I taught. Yes, that must be it. Maybe she'd worked as a peer tutor in one of my classes.

Well, however I knew her, I hoped all was going well with her in her young life.

One hazy afternoon half-way through August I went shopping at Wal-Mart. Pushing the cart filled with just-paid-for household goods, I was on the homestretch, scrounging for car keys in my purse and aiming the cart through the exit electronic monitors.

But there were bodies blocking my way, and when I looked up, THERE SHE WAS! Along with an older woman (her mom?) and another teen who might have been her sister.

I called out, "How *are* you? How've you *been?*"

Talk about deer eyes in the headlights. To make a short story shorter, she'd never seen me before in her life. She'd never been a student at the high school where I taught nor in the few moments we visited could we find any other commonality.

So how did I know her? Why was she so much on my mind? Had I had a brush with an alternate reality where she *had* attended my high school? Had her doppelganger at some point in some way passed through my life just long enough to leave an impression? Or had all of that teaching finally produced mushy brain fruit?

Alternate Reality TWIDDERS often center on doppelgangers. For our purposes, we'll define a doppelganger as a double of a person; not just a dead-ringer look-alike but the real McCoy. Incidents involving a double are often referred to as a *bilocation*.

Like life isn't confusing enough already!

THE SISTER WHO WASN'T THERE

Sandra Lee K. (2008) was about 14 years old when she experienced an Alternate Reality TWIDDER.

My father was working one night (fire fighter). My sister, seven years older, was out. My mother and I were home alone.

Mom had locked up downstairs and was in the bathroom on the second floor. I was looking for my slippers and remembered I'd left them in my sister's room in the attic earlier that day.

The attic was one room and had a half wall along the staircase instead of a railing. [My sister] had a small lamp attached to the top of that, and to turn on the lamp, there was a string hanging down with a metal washer tied to it so it could be turned on from the bottom of the stairs.

The first three steps were wedge-shaped and the rest above that were regular steps. I pulled the string and turned on the light. Then I started walking up the stairs. For some reason, I don't know why, I leaned forward a few steps from the top and peeked around the half wall.

There across the room was my sister's single bed. She was standing on the other side of it. Her flannel nightie with blue flowers on it was lying on the bed.

She picked it up and put it over her head and pulled it down over herself. Then she pulled her long hair out of the back, pulled the collar out, and began to button the three buttons in the front.

That's when I coughed, and my sister threw her hands up like I'd scared her half to death, and then she DISAPPEARED!

I screamed and ran downstairs to the second floor. My mother met me and asked why I'd screamed. I told her! We went back up to the attic together, and Mom looked under the bed. There was no place to hide up there not even a closet. Mom was afraid it meant that something awful happened to my sister, but she came home later and was just fine.

I'm now 60. I've always wondered why I saw what I did. It wasn't just a flash of something, which could have been imaginary. I'd watched my sister getting ready for bed, which was a pretty normal thing.

Sandra finished her narrative with the plea, "Do you have any ideas about what this may have meant?"

THE TESTING LABORATORY

"Nermal" (Like many others who aren't sure they want anyone to know their names but still want their stories to be heard) wrote about the following incident that had occurred during the late 1960's in central Florida at the testing laboratory where "Nermal" worked.

Two of us witnessed this event: Gene Z and myself.

The two of us were in a small laboratory discussing the expected move to a much larger building. This room was about 25 feet long by six feet wide (desks and work benches were along both walls), and it was only wide enough for one person to walk through. At the end of the room to the right was an even smaller storage closet.

The two of us were sitting diagonal from each other when we saw Eloise S. walk between the two of us and enter the storage room (about four feet by six feet).

We looked at each other and Gene Z. remarked, 'That was Eloise!'

After a few minutes when she did not come out, I looked into the storage room. No Eloise! This room had a sink and exhaust fan at one end and a small storage cabinet on the other. There was no other exit except back the way she entered.

Thinking that something must have happened to Eloise, I went over to the X-ray laboratory where she was working. She had been in the darkroom most of the morning but was wearing the same blue dress that we saw earlier.

I asked her if she was okay, and I told her that we saw someone that looked like her walk by and enter the storage room. She indicated that nothing out of the ordinary happened all morning."

HIDE AND SEEK

Robert F. (2008) wrote about a bilocation experience involving his younger sister. It was late spring or early summer in 1992. Robert and his sister (eleven and nine years old, respectively) were living with their parents in a house in rural south-central Pennsylvania.

> She and I had just returned home from the store. It was the early afternoon on a Saturday, and the day was uncommonly misty and cool. I challenged my sister to a game of hide and seek. She agreed as long as she could go back in the house at 2:00 P.M. to watch her favorite cartoon show on television. I looked at my watch; it was 12:30 P.M., so I figured we had plenty of time to play a few games.

> She decided to hide first, so I went to the front door of our house and counted to ten. 'Ready or not, here I come,' I shouted and started around the front of the house toward the carport on the north side of the house.

> My parents had just bought a new refrigerator a couple days before and had left the large box it came in out in the carport. I figured my sister would want to hide in there. I quickly rounded the corner to the carport and looked at the box laying on the floor. I saw a shock of blonde hair enter the box from the open side, which was opposite; the box even moved a little. I laughed, knowing that I had caught her so easily.

I ran over to the box and looked inside. The box was empty! Perplexed, I stood around for a minute, wondering how I could have missed her. Had she been there at all? I reasoned that I had just imagined seeing her and continued walking through the carport toward the backyard.

I stood there scanning the spacious backyard for a few moments, thinking of where to look next. I returned my gaze to the middle of the yard and saw my sister—plain as day—standing about 50 feet from me, smiling in her typical way. Then she waved at me! I thought that she had given up the game because she wanted to go in the house. Strangely, she was wearing a yellow t-shirt, different than what she had been wearing when we started playing.

Something was strange about her. Somehow, I was afraid to go tag her. I started walking out to talk to her. I must have looked down briefly, but when I looked up again, she was gone. The closest thing for her to hide behind would have been a couple of apple trees standing about another 50 feet behind her. I knew it was impossible for her to have run back there that fast.

I was amazed—and very scared. I got a chill. Something was not right. Was I seeing things? I literally stood there for a few minutes trying to make sense of it all. Twice she had been there; twice I was wrong. At this point I was feeling impatient to find her (and anxious, I imagine, to prove that I wasn't losing my mind!) So I walked around to the south end of the house where a large maple tree stood. In our frequent hide and seek games, I would often hide up in the limbs of that tree, but either way, I was running out of places to look.

I walked around the corner and looked at the tree, then scanned the area around it. Looking back at the tree, I saw someone standing behind the tree trunk. Whoever it was wore a purple shirt, which stuck out the side from behind the tree. Taking care not to look away for one second, I fixed my gaze on the tree and ran at it. I went behind the tree, fully expecting to find her.

She was not there. I looked frantically all around the area, even in some shrubs at the very edge of the yard, to no avail.

At this point, I went back in the house to tell my mother that I couldn't find my sister and that I thought something strange was going on.

And there, on the living room couch, sat my sister, wrapped in a shower robe with a towel around her head, casually watching TV. I gasped in disbelief and asked her where she had been.

She said that she had been hiding in the box, but I never came by, so she went to the backyard and, not seeing me at all, she hid behind the maple tree for awhile.

She told me that she must have waited an hour without seeing me, so she got tired and went inside, almost expecting to see ME inside as well. And that she had taken a shower before sitting down to watch her cartoon.

It was already after 2:00 P.M.! I told my sister that I had seen her wearing different shirts that she indeed owned. She thought it was strange since she had just been thinking earlier about what to wear when she got out of the shower.

THE BOYFRIEND

Leslie R. (2009) wrote about a doppelganger episode involving her boyfriend from spring 2009.

> I was getting ready for bed around 1 A.M. as we're [my boyfriend and I] both night owls. My boyfriend told me he was in the middle of an art project and would be in bed when he finished up. This was nothing unusual as he is a graphic designer and draws as a hobby. I kissed him goodnight and climbed into bed.
>
> I woke up a few hours later and rolled over and saw him in bed next to me, sleeping like he usually does on his stomach with his arms up under his pillow. It seemed odd at the time, but I felt an urge to check the time but pushed it down as I snuggled up next to him.
>
> I also found it odd that he had come to bed without shutting off all of the lights in the house like he normally does. I could also still hear the menu screen from the DVD we had been watching playing, but I was tired, so I put my arm around him, found his hand, [and] snuggled into him.
>
> The next morning, I woke to my alarm on my left side. (I only sleep on my left side when he is in bed with me; when I'm alone, I sleep on my right, facing the wall. But he wasn't in bed.)
>
> I got up to see where he was because he never gets up that early and found him asleep on the couch [with] the TV still on the menu screen for the same DVD. [He was] in the clothes he had been wearing the night before when I went to bed. He had a half-completed drawing under him.

Still not thinking anything was off, I woke him and asked him why he got out of bed, dressed, and came to the couch to sleep.

He got a confused look on his face and said that he had never come to bed the night before; he just fell asleep where he was. I have tried to tell myself every day since that I was dreaming, but there was no way that was a dream.

I could feel him, see him, and smell him; and I could hear the DVD from the living room.

THE DAUGHTERS

In March of 2009 Kathy B. wrote about two doppelganger experiences with her daughters. Kathy, a wife and mother, was living in a farmhouse in rural Wisconsin at the time of the TWIDDERS.

I live with my husband, aunt, and four children. The first instance [of bilocation] was at night in 2004. My daughter had gone to a birthday celebration about 45 minutes away. The next day I had planned to take her with me to buy some livestock in southern Illinois to start my business.

Around ten in the evening, I got ready for bed. My husband had already gone to bed about 20 or so minutes ahead of me. By the time I got in the bed, he was snoring as he usually goes off to sleep as soon as his head hits the pillow.

I am not so lucky and it takes me awhile to fall asleep. I lay in the bed thinking about how the day was going to go off tomorrow. I'd say I'd been in the bed five minutes.

Then the door opened up to my room. The hall light was on, and I could see clearly it was my daughter, Jessica. *She must have gotten back from the party*, I thought. She stood in the doorway for awhile staring at me.

She was a solid form just like anybody is and not a ghost, but I found it odd she wasn't speaking to me and had a tint of bluishness in her face, but [I] thought it was just the darkness of my room reflecting on her complexion. She had a bland expression on her face, but her eyes were wide, just staring.

I decided, well, I'll talk to her, and I said, 'It's late. You better get some sleep. We have a big day tomorrow.'

She responded and said, 'I know,' and closed the door and left.

Then as I lay thinking, I thought there were a few things more I wanted to say to her before she went to bed.

I got up right away only to find that she was nowhere in the house. At that point, I was very angry, thinking she had gone out again. I called her cell number and she picked up, surprised by my anger.

She insisted that she had never been home and was still at the birthday party in Janesville. I didn't believe her and asked to speak to Mrs. Davis, who was the mother of the child having the party in her home.

She came on the cell and confirmed that Jessica was there and had been there the entire evening. I asked Mrs. Davis if I could talk to her on her home land line phone. So immediately I called their land line, and Jessica picked up.

I was floored. How could it be? She was for certain no doubt in Janesville 45 minutes away, yet I saw her in my bedroom doorway no more than ten minutes earlier. It is a complete impossibility. I could not explain it.

The second instance happened to my aunt in March 2009. She was looking out her window and noticed my other daughter, Caroline, walking outside without her hat on.

She said her hair was back with a clip in the back, and she was wearing a grayish-blue coat. She said she was no more than ten feet from the window. She walked slowly and serenely past like she was really enjoying the outdoors. This was in the morning.

I was nearby in the next room when my aunt said that Caroline was outside, and she didn't have a hat on. I said that was odd that she would be outside already. All of my children sleep in on the weekend usually. I wasn't too concerned about it and walked to the other side of the house and started watching TV, then after that, got myself ready to go outside to do the farm chores.

Caroline then came down the stairs! She was in her bed clothes and groggy. I asked her what she was doing outside earlier, and she looked surprised and said she had just woken up and had not been outside at all.

I found it odd that she does not own a coat that is grayish-blue [like] my aunt saw. Caroline is a little girl with red hair. She is very hard to miss. I thought maybe my aunt saw someone else, but on the farm we don't have neighbors nearby, and none that have red hair. My aunt insisted that it was definitely Caroline. She is dead sure.

I can't explain that one either, unless Caroline sleep-walked, got herself ready for the day, and then went outside with someone else's coat. Since it was morning, I can't accept that either. I cannot explain it.

She is not old enough (nine) for sneaky teen stuff and is a self-conscious, shy girl. Also, I had never told my aunt about the *Jessica incident* but shared it with her after this happened to her.

ROSIE

In December of 2008 Paul M. wrote about a TWIDDER with his dog.

This incident took place at 7 A.M. on October 16th, 2008 at my home in Tamworth, Staffordshire, England.

I got up at my usual time of 7:00 A.M. and followed my usual routine of making a cup of tea and letting my two dogs outside into the garden.

Polly, my small terrier, disappeared into the back garden, and Rosie, my German Shepard, took up her position on the corner of the yard between the back garden and the back door.

She was in view to me from this position. Polly returned from the garden, and I let her into the house. Rosie, as usual, suddenly shot off into the back garden (which is out of view from my position at the back door).

I knew straight away she was running down to the bird table at the bottom of the garden to chase the pigeons that always visit at this time.

I turned around to go back inside to finish making my cup of tea. I could still hear Rosie barking. As I opened the door, I no longer could hear the barking and just assumed the pigeons had gone. I entered the house and—you could have knocked me down with a feather—lying down on her bed was Rosie!

I just don't know what happened. I have no theories. I am just totally baffled and would love someone to try and explain this. I know there will be people who say I hadn't noticed her enter the house or that I had let her back in and then forgot, but I can assure you this is not the case.

I am 110% confident that it happened exactly as I have related here (p. 21).

THE GARDEN

James (2007), a lover of gardening, had a very unique gardening experience. When James had moved into a three bedroom house, complete with a 150 foot garden, he hardly made time for anything else but gardening.

After planting some new borders, I proceeded to my den for a quick cup of tea and a snack. The den itself was very small and had one window and a small gap near the roof for spying out of.

When I proceeded to take a peek out of this hole, I was amazed at what I saw. There was a small pond and beautiful palm trees surrounding it.

The dull, light green lawn of my garden was replaced with lush, dark green grass, and it was tremendously sunny! The shed was already on the property when I bought it, but the previous owner—a middle-aged man who never went in there—said he had no need for it, so he left it. It was about ten to twenty years old and in great condition.

Could it be that that shed is some sort of portal, depicting another place on this Earth? The garden itself had not changed when I took a look out the normal window.

Unfortunately, I haven't been able to take a look [again] at that paradise-like place even when I look out the gap . . . although I'd be eager to see if people on the other side can see me!

DOUBLE DAD

A young man who chooses to remain anonymous (2009) wrote about an enigma he experienced.

I saw my dad's doppelganger. I live[d] with my parents and always [wanted] to get away for a while to have time to myself.

I had been hanging out with a friend. After I dropped him off, I went to a graveyard. I [went] there to enjoy the silence of the night, but I kind of got a creepy feeling and left soon. I went to get some fast food before I went home.

When I was turning at the stoplight to head home, I looked right over at the car next to me; it was my dad in his car. He had a very tired look on his face.

I was wondering why he didn't wave at me or act like he saw me. I also wondered why he didn't have his hat on. He never leaves the house without wearing one. I also wondered what the heck he was doing out that late anyway because he [was] usually in bed or getting ready to go to bed. I thought I saw someone in the car with him and thought my mom must have been with him. I just assumed they were going to rent a movie or something.

So I got home and there was my dad's car in the driveway. I walked in and he and my mom were watching TV. They said they never left. They had been there all night.

ME, MYSELF, AND I!

Steven M. (2009), too, had a very personal doppelganger incident in mid-November 2008.

I lived in a very little town named Palatka, Florida. There is not much to this town, but what happened to me blew me away. I thought maybe my friends were playing a sick game with me or maybe they were drunk, but what they showed me was even more impossible for my mind to take in.

As I was talking to my friends on my cell phone from my lunch break at work, they said I just went into a bar where they were. I told them I was at work, but they didn't believe me.

So they decided to follow the person they said was me and take a cell phone picture. Sure enough, if you would have held a mirror up to me, the picture they showed me was shocking, because 'I' was sitting at the bar waving at them with a beer in my hand.

Then it really got weird. My friend was sitting there and said to me, if 'I' am in front of them now, then how am I speaking to them on the phone? I told them the same thing, but by the time they turned back around to the bar area, they said 'I' was gone—but 'I' left the barkeep a tip.

They tried to find out what happened, but the barkeep stated that he never saw *me* leave the bar that night. To this day, my friends are still freaked out by that.

Chapter 10.

Uh Huh

Betwixt dialogues on time/no time, wormholes, doppelgangers, branes, and probability strands, what say we muddy the TWIDDER waters even more?

TWIDDERS appear to be spontaneous; the experiencers have no control over the event itself. They never expected it, were shocked or befuddled or floored or amazed or fascinated or worried that they were going nuts, or worried that others would think they were going nuts, or all of the above when it happened, and certainly had no control over experiencing one again.

For the most part. Evidently.

Yet we have Toynbee and his repeated time-slips to the past.

NOTE: At this point in time (or non-time?) I'm making the HUGE assumption that all of the experiences recorded in this volume truly happened. That the experiencers who have so kindly, often at the risk of ridicule, shared these happenings are neither delusional nor fiction writers hoping to dupe readers.

That said, there are certain time-slip accounts—most especially the story entitled "The Church" in Chapter 2 that describes a visit to a rapidly aging chapel—that may be suspicious. I came across similar scenarios always involving rural structures of some sort where the visitors all supposedly experienced the rapidly-aging-structure-syndrome. These additional "accounts" I chose not to include in this book because they struck me as fictional. I also came across critics who decry this form of TWIDDER as an urban (or rural, in this case) legend at its silliest and believe that the reader who takes it as gospel is a prime candidate for investing with Bernie Madoff.

However that may be and if one places faith in the verity of the time-slips in this volume, are there triggers that may increase the odds of experiencing a TWIDDER?

For Toynbee, was it his intense, hyper-focus on historic events at the very sites he was visiting that facilitated the TWIDDERS?

Triggers that may enable a TWIDDER might include these conditions:

> Hyper-focus/prior specialized knowledge on the part of the experiencer (such as Toynbee brought to the geographic sites where he experienced time-slips);
> Locations drenched in emotion/violence (Civil War battlegrounds, etc.);
> Electrical storms/electromagnetism at work;
> Psychometry (for time-slips, touching an object and seeing a scene from its past).

HYPER-FOCUS/PRIOR SPECIALIZED KNOWLEDGE

Toynbee is the classic example of a TWIDDER experiencer who had totally immersed himself in the history of the sites where he experienced time-slips.

In Chapter 3 aviator Martin Caidin (1994) shared "The Case of the Curious Shell Casing" He and a crew of long-time buddies were flying an old Boeing B-17G Flying Fortress from Arizona to England.

While Bert Perlmutter napped on a wall-mounted bench in the fuselage, his comrades watched, drop-mouthed, a flight scene from World War II. They viewed a taut battle as two gunners hammered away at the enemy, shell casings bouncing around the interior of the plane.

They saw one gunner lurch back, shot, and the other grab him, shepherding him to a freezing air stream to staunch the blood gushing from his severed hand.

Seventeen years earlier, the now-slumbering Perlmutter *was* that shepherding crewmember who saved the other soldier's life!

Caidin noted in hindsight that Perlmutter had been sleeping restlessly, as though having a nightmare. Had he been reliving that horrific experience and by infusing the old airship with his unconscious but intense memories, caused a time-slip?

Author Joan Forman (in Fanthorpe's *The World's Greatest Unsolved Mysteries,* 1997) has a different take on this time-slip trigger. She believes that if a person is interested in but not concentrating on his/her surroundings, a time-slip may occur. She cites her own experience in Derbyshire, England, while visiting Haddon Hall. Pausing to admire the architecture, she saw children from the 1600's at play on the stairs. (See Chapter 3 "Haddon Hall.")

Caveat. While TWIDDERS may be enabled by hyper-focus, or prior knowledge of historic occurrences at a particular site, or even by having-an-interest-in-but-not-really-concentrating-on an historic site, this book recounts many occurrences wherein the experiencers had NO prior knowledge NOR any particular interest in the spot where their TWIDDER occurred. Go figure.

LOCATIONS DRENCHED IN EMOTION/VIOLENCE

Do we have a clue what happens to the very nature of a field like Antietam, the site of the one-day bloodiest battle of the American Civil War with its 23,000 recorded casualties?

When such intense horror, sorrow, pain, anger, fear, death, and regret on the part of thousands of humans are spilt at one location in a relatively short period of time, are the very atoms of the site infused with the memory of the event?

Or, in some way we obviously don't yet comprehend, is it possible for the event itself to create a physical bond that exponentially increases the odds that a visitor will experience a TWIDDER at that site?

There's a fascinating little volume entitled *The Hidden Messages in Water* by Japanese research scientist Masaru Emoto (2004). Using high-speed photography, Dr. Emoto discovered that crystals formed in frozen water reveal changes when specific, concentrated thoughts are directed toward them.

Think LOVE as the water crystals form, and the photos reveal breathtakingly symmetrical, intricate, sparkling crystals. Think HATE and a crystal cannot even form; instead, the pictures document grotesque, malformed blobs of frozen water. Water crystals are transient; warmth dissipates even the most glorious crystal in seconds.

What if, on a quantum level, thoughts affect ALL matter, not just water? What if they permanently affect the make-up of the molecules of a tree that will yet live 300 years?

Could that concept be the first baby step in understanding why TWIDDERS more easily occur in locations drenched in emotions/violence?

Uh huh, I, for one, think so.

And yet, there are other locales—such as Bold Street in London, England, where life has forever been mundane. But it was here that police officer Frank walked into a defunct store (Chapter 3 "The Shop That Wasn't"), and other Brits have also experienced time-slips.

For example, a construction worker renovating the Lyceum building on Bold Street, whose watch ran *backwards* for two hours one day. Another day, he dropped his safety helmet, and when he looked down a few seconds later, it had vanished, yet no one was within 50 feet of him.

Oh heck, here it comes again, the dreaded . . .

Caveat. Most visitors to historic sites don't have a TWIDDER. But I've never read of a time-slip occurring at the local McDonald's whereas I've read many accounts from battlefield visits. Surely if you want to enhance the odds of experiencing a TWIDDER, a visit to a battlefield would be more conducive than a trip to Safeway.

ELECTRICAL STORMS / ELECTROMAGNETISM AT WORK

Prolific science author and professor of biochemistry, Isaac Asimov, once defined animate versus inanimate as that which makes an effort. (Hmmm, that definition renders a number of students I've had over the years as the walking dead.)

In physics, motion (making an effort) requires force and energy. Energy is the ability to do work. Forces cause motion (acceleration). As we know from "Gregg Eating those disgusting Watermelon Sandwiches," (Chapter 1) electromagnetism is one of the basic forces of our universe.

To undergo a TWIDDER *something* happens physically, which means that *something* is making an effort, expending energy, and being worked on by one of the four basic universal forces. Could environmental conditions involving electricity be a trigger for a TWIDDER?

Many TWIDDER experiencers describe electrical phenomena at the beginning and end of their time-slips. The British ladies saw a century old version of Versailles when they were visiting France during a summer that had been plagued by electrical storms.

Keith Manies (Chapter 2 "The Brave and the Buffalo") described an electrical buzz or whine that he heard. Later that evening the area was hit by one of the most intense lightning storms he'd ever seen. And being from Kansas' tornado alley, he'd seen a LOT!

Other experiencers talked about a sense of being out-of-sync, being enveloped in a yellowish fog, or feeling like trying to move through gelatin. It seems that many experiencers used their physical senses to try to understand the changes in the material world surrounding them before, during, and/or after their TWIDDER encounters.

Something happens physically when a TWIDDER hits. The mechanics of how a time-slip initiates is a big mystery. But it doesn't happen in a vacuum. Electricity may be one of the basic forces that activates the TWIDDER switch.

Sorry, but I've gotta do it . . .

Caveat. While being vitally interested in having a trip-to-the-past TWIDDER, I have NO interest in triggering a time-slip by standing tall in the middle of a wheat field during a violent thunderstorm. (Should I ever be desperate enough to do so, odds are I'd more likely have a Near Death Experience!)

Should you choose to attempt a TWIDDER by seeking out lightning, you do so at your own risk!

PSYCHOMETRY

In 1842 Joseph Buchanan coined the term "psychometry." For a brief explanation, see Wikipedia. Buchanan believed that all things give off an emanation. He wrote, "The past is entombed in the present; the world is its own enduring monument."

He theorized that a person with enough sensitivity could hold an object in his/her hands and obtain paranormal information about the object or its owner. In the case of a researcher hoping for a TWIDDER, he/she would be seeking an actual trip to the object's past.

Alice Pollock toured Leeds Castle in Kent, England. While in Henry VIII's rooms, she touched objects in an attempt to trigger a time-slip. After a period of receiving no impressions, the room suddenly changed.

Instead of the modern, comfortable furnishings and finishes, the room was cold and bare. The carpet was gone, and logs were now burning in the fireplace.

A tall woman in a white dress was pacing the room, her face in deep concentration. Shortly thereafter, the room returned to its modern state. The rooms had been the prison of Queen Joan of Navarre, Henry V's stepmother, who had been accused of witchcraft by her husband.

One final caveat. Whether we go for a post-graduate degree in history or invest in New Age psychometry courses, there's no guarantee of having a TWIDDER. No matter how much we employ triggers trying to experience a time-slip, we may never be able to join the exclusive TWIDDER club. On the other hand, nothing ventured; nothing gained!

Well, we've come full circle. This book started with an account of a TWIDDER and an attempt through scientific theory to come to terms with how one could occur.

We've ended with a TWIDDER, and I'm still looking for carved-in-stone confirmation that time-slips are a factual and natural part of mortality. Nuts or not, I've come across testimony from Near Death Experiences(ers) (NDE'rs) that support current physics concepts.

NDE'r Kimberly Clark Sharp explained in her autobiography, *After the Light: What I Discovered on the Other Side of Life* (1995), that when she was out of her physical body, time didn't exist as we know it. Sharp noted that "Instead of past and future, there is a profound sense of *now*."

Uh huh. Remember physicist Julian Barbour's theory of the great Now?

In "Journey Through the Light and Back," NDE'r Mellen-Thomas Benedict wrote, "Scientists perceive the Big Bang as a single event which created the universe. I saw during my life after death experience that the Big Bang is only one of an infinite number of Big Bangs creating universes endlessly and simultaneously. The only images that even come close in human terms would be those created by super computers using fractal geometry equations."

Score one for physicist Michio Kaku!

Whether the *proof* comes from state-of-the-art physics theory or reports from survivors of death, at this point in time TWIDDERS are still classified as metaphysics, not mainstream science.

As my seventeen year old would say, "Whatever."

Do I think time-slips occur? Uh huh.

Do I understand how they can occur? Huh uh.

Would I like to experience a trip-to-the-past? You bet!

Do I envy those who have? Oh, yeah.

For those of you who, like me, hope to one day stumble through a porthole into another time or another place, may the force (electrical?) be with us!

References

45 minutes vanished. (2008, October). Retrieved from http://paranormal. about.com/library/blstory_october08_02.htm/

A blip in reality. (2002, May). Retrieved from http://paranormal.about.com/od/ timeanddimensiontravel/a/aa012306.htm

A brief leap in time. (2004, July). Retrieved from http://paranormal.about.com/ library/blstory_july 04_16.htm/

Alvarez, R. (1998, January 28). *Mystery meadow*. Retrieved from http://www.ufofreeparanormal.com/stories/viewstory.php ?sid=645.htm/

Amazing pterosaur sighting. (2002, August). Retrieved from http://paranormal.about.com/library/blstory_august02_10. htm/

Another fog space-time slip. (2007, July). Retrieved from http://paranormal.about.com/library/blstory_july07_10.htm/

Around the house. (2006, May 17). Retrieved from http://www.ufofreeparanormal.com/stories/viewstory.php ?sid=1171/

Asher, R. (2009, September 03). *Missing time in New York*. Retrieved from http://paranormal.about.com/od/timeanddimensiontravel/ a/tales_09_03_09t.htm/

Baby monitor time warp. (2009, May). Retrieved from http://paranormal.about.com/od/timeanddimensiontravel/ a/tales_09_05_21t.htm/

Bailey, D. (1995, June 12). *Bar boys*. Retrieved from http://www.ufofreeparanormal.com/stories/viewstory.php?sid=44.htm/

Benedict, M-T. (n.d.). *Journey through the light and back.* Retrieved from http://mellen-thomas.com/stories.htm/

Berke, A. (2008, April). *Wormhole on a country road.* Retrieved from http://paranormal.about.com/library/blstory_april08_12.htm, April 2008.htm/

Bowen, C. (1980). *Encounter cases from flying saucer review.* New York: Signet Books.

Bratton, R. (2004, September). *Instant replay.* Retrieved from http://paranormal.about.com/od/timeanddimensiontravel/a/aa012306_2.htm/

Brennan, J. H. (2002). *Time travel: A new perspective.* Woodbury: Llewellyn.

Budd, D. (n.d.). *The dragon's triangle.* Retrieved from http://www.bella online.com/articles/art59502.asp/

Burgoyne, T. (2006, July 20). *Time slips . . . ?* Retrieved from http://www.unexplained-mysteries.com/forum/lofiversion/index.php/t74556.html/

Caidin, M. (1994). *Ghosts of the air: True stories of aerial hauntings.* New York: Galde Press.

Clark Sharp, K. (1995). *After the light: What I discovered on the other side of life that can change your life.* New York: William Morrow and Company.

Conway, V. (1999). Exit, stage left: Audience from the past gives music student stage fright. In J. Spencer (Ed.) *The encyclopedia of the world's greatest unsolved mysteries.* Fort Lee: Barricade Books.

Dad's doppelganger. (2009, March). Retrieved from http://paranormal.about.com/od/humanenigmas/a/tales_09_03_12t.htm/

Daughter's bilocation. (2009, March). Retrieved from http://paranormal.about.com/od/humanenigmas/a/tales_09_03_15t.htm/

Dimension shifts in Yorkshire. (2009, May). Retrieved from http://paranormal.about.com/od/timeanddimensiontravel/a/tales_09_05_01t.htm/

Disappearing hole. (2003, December). Retrieved from http://paranormal.about.com/library/blstory_december03_25.htm/

Disappearing, appearing trees. (2008, September). Retrieved from http://paranormal.about.com/library/blstory_september08_27.htm/

Distorted time. (2004, April 28). Retrieved from http://paranormal.about.com/b/2004/04/28/distorted-time.htm/

Dog mystery. (2008, December). Retrieved from http://paranormal.about.com/library/blstory_december08_21.htm/

Doppelganger time warp. (2008, August). Retrieved from http://paranormal.about.com/b/2008/08/12/doppelganger-time-warp.htm/

Fanthorpe, L. (1997). *The world's greatest unsolved mysteries*. Toronto: Hounslow Press.

Faulks, N. (2008, August). *Disappearing glasses*. Retrieved from http://paranormal.about.com/library/blstory_august08_06. htm/

Folger, T. (2007). Newsflash: Time may not exist. *Discover*, Retrieved from http://discovermagazine.com/2007/jun/in-no-time/article_view? b_start:int=1&-C/

Freaky doppelganger experience. (2009, March). Retrieved from http://paranormal.about.com/od/humanenigmas/a/tales_0 9_03_04t.htm/

Future city. (2005, September). Retrieved from http://paranormal.about.com/library/blstory_september05 _11.htm/

Gained time. (2008, August). Retrieved from http://paranormal.about.com/library/blstory_august08_26. htm/

Garden of the fifth dimension. (2007, July). Retrieved from http://paranormal.about.com/od/timeanddimensiontravel/ a/aa072307_2.htm/

Gettysburg. (1981). Retrieved from http://cc@mindspring.com/

Ghost in a hurry. (2008, December). Retrieved from http://paranormal.about.com/library/blstory_december08_08.htm/

Ghost of Little Big Horn. (2009, June). Retrieved from http://paranormal.about.com/od/hauntedplaces/a/tales_09_06_17t.htm/

Ghosts of Hawaii. (1995, March). Retrieved from http://www.faqs.org/faqs/folklore/ghost-stories.htm/

Giant raptor witnessed by 5. (2005, January). Retrieved from http://paranormal.about.com/od/othercreatures/a/tales_09_01_19t.htm/

Gilbert, S. (2006, June 2). *Disappearing hallway.* Retrieved from http://www.ufofreeparanormal.com/stories/viewstory.php?sid=1205.htm/

Gisby, L. & C, Simpson, G. & P. (1989). We'll leave the speed of light on for you: 20th-century guests stay in 19th-century hotel. In C. Berlitz (Ed.) *World of strange phenomena.* New York: Fawcett.

Goonay, Z. (2006, October-December). The essence of time. *The Fountain Magazine,* 14(56). Retrieved from http://www.fountainmagazine.com/article.php?ARTICLE ID=782/

Griff. (2008, August). *Pittsburgh thunderbird.* Retrieved from http://paranormal.about.com/library/blstory_august08_17.htm/

Guth, A. (2001, October 7). *Inflation and the accelerating universe*. Retrieved from http://insti.physics.sunysb.edu/~sterman/OWPprogram.html/

harvestmom. (1999, March 28). *House in the woods*. Retrieved from http://www.ufofreeparanormal.com/stories/viewstory.php?sid=163.htm/

Hensnchicks, I. (2009, June). *Time glitch alters fate?* Retrieved from http://paranormal.about.com/od/timeanddimensiontravel/a/tales_09_06_20t.htm/

Hill, E. (1997). Up in alms. In C. Kenner & C. Miller (Eds.) *Strange but true* (pp. 35-36). Woodbury: Llewellyn.

Holmgren, J. (2008, December). *Missing time near area 51*. Retrieved from http://paranormal.about.com/library/blstory_decemb/

In bed with a doppelganger. (2009, May). Retrieved from http://paranormal.about.com/od/humanenigmas/a/tales_09_05_07t.htm/

Jamafan. (2003, June). *Time slip to d-day*. Retrieved from http://paranormal.about.com/library/blstory_june03_09.htm/

Johnson, S. (n.d.). *Lieutenant Colonel Frank Hopkins*. Retrieved from http://www.mercuryrapids.co.uk/articles5.htm#ufofiles-pacificbermudatriangle01122006/

Journey to the past. (n.d.). Retrieved from www.paranormalabout.com/od/timedimensiontravel/a/aa 012306.htm/

Jump in time. (2006, May). Retrieved from http://paranormal.about.com/library/blstory_may06_02.htm/

Jung, J. *Memories, dreams, reflections.* (1965). New York: Random House.

Kaku, M. (2009). *Physics of the impossible.* Norwell: Anchor Press.

Kenner, M. (1997). *Strange but true.* Woodbury: Llewellyn Publications.

Lee, M. (1995, March 31). *I.e. the bowling ball.* Retrieved from http://www.ufofreeparanormal.com/stories/viewstory.php ?sid=475.htm/

London time slip. (2009, May). Retrieved from http://paranormal.about.com/od/timeanddimensiontravel/ Time_and_Dimension_Travel.htm/

Lowe, S. (2007, July). *A time inconsistency.* Retrieved from http://paranormal.about.com/library/blstory_july07_20.htm/

Manies, K. (2008, August). *Honest Injun time slip.* Retrieved from http://www.forteantimes.com/forum/viewtopic.php?p=18 3165&sid=dec37fca8ff9 8b1d1518a8aeb81f41cf/

Man's spirit saved future self. (2008, August). Retrieved from http://paranormal.about.com/library/blstory_august08_29. htm/

Missing time and out of place. (2004, November). Retrieved from http://paranormal.about.com/library/

Missing time. (2003, March). Retrieved from http://paranormal.about.com/library/blstory_march03_25. htm/

MOMDBOM9, I. (2009, May). *Husband's bilocation.* Retrieved from http://paranormal.about.com/od/humanenigmas/a/tales_0 9_01_06t.htm/

Moore, T. L. (1999, March). *The burned farmhouse.* Retrieved from tlm887@mail.usask.ca/

Mystery caboose. (2008, January). Retrieved from http://paranormal.about.com/library/blstory_january08_10 .htm/

Nermal. (2009, February). *Lab doppelganger.* Retrieved from http://paranormal.about.com/od/humanenigmas/a/tales_0 9_02_18t.htm/

Nettdan. (2001, October). *Time stood still.* Retrieved from http://paranormal.about.com/library/blstory_october01_1 3.htm/

O'Neill. (2003, October 7). *Time slip in Australia.* Retrieved from http://paranormal.about.com/b/2003/10/07/time-slip-in-australia.htm/

OPFOR1. (1997, September 26). *Ghosts of the battle of Sharpsburg.* Retrieved from OPFOR1@aol.com/

Owens, C. (2004, November). *The phantom travel agency.* Retrieved from http://paranormal.about.com/library/blstory_november04_22.htm/

Path to . . . where? (2008, December). Retrieved from http://paranormal.about.com/library/blstory_december08-10.htm/

Phantom ambulance. (2009, February). Retrieved from http://paranormal.about.com/od/timeanddimensiontravel/a/tales_09_02_20t.htm/

Phillips, Y. (2009, April). *Raptor sighting in Georgia.* Retrieved from http://paranormal.about.com/od/livingdinosaurs/a/tales_09_04_15t.htm/

Physics class. (1994, June 16). Retrieved from http://groups.google.com/group/alt.dreams/search?group=alt.dreams&q=physics+class&qt_g=Search+this+group/

Polymathicus, I. (2007, September 30). *The timeless universe of Julian Barbour.* Retrieved from http://polymathicus.blogspot.com/2007/09/timeless-universe-of-julian-barbour.html/

Psychometry (paranormal). Wikipedia. Retrieved (2009, June 15) from http://en.wikipedia.org/wiki/Psychometry_%28paranormal%29/

Pullman, P. (1995). *The golden compass.* New York: Ballantine Books.

Randall, L. (2005). *Warped passages: Unraveling the mysteries of the universe's hidden dimensions*. Hopewell: Ecco Press.

Randles, J. (2000). *Time storms*. London: Berkley.

Retro. (2005, March). *Lost time at the movies*. Retrieved from http://paranormal.about.com/library/blstory_march05_23. htm/

Ryan, A. (2006, April 20). *Missing mansion*. Retrieved from http://www.ufofreeparanormal.com/stories/viewstory.php ?sid=899.htm/

Same day twice. (2003, October). Retrieved from http://paranormal.about.com/library/blstory_october03_1 9.htm/

Simmons, J. (1998, August 7). *The phantom street*. Retrieved from http://www.ufofreeparanormal.com/stories/viewstory.php ?sid=574.htm/

Slemen, T. (2000). *Haunted Cheshire*. Liverpool: The Bluecoat Press.

Slemen, T. (n.d.). *Merseyside timeslips*. Retrieved from http://www.geocities.comArea51/Capsule/1851/timeslips. html/

Squirrel. (1981). Stationery moment in time: Man buys envelopes from Victorian-era shop clerk. In C. Wilson & J. Grant (Eds.), *The directory of possibilities*. London: Webb & Bower.

Stravato, L. (2006, July). *Missing time in the kitchen.* Retrieved from
http://paranormal.about.com/library/blstory_july06_06.htm/

Stumbled into the future. (2008, October). Retrieved from
http://paranormal.about.com/library/blstory_october08_0
3.htm/

Swartz, T. (n.d.). *On the edge of time: The mystery of time slips.*
Retrieved from http:uforeview.tripod.com/timeslips.html/

Tallberg, P. (2002, September). *The vanishing church.* Retrieved
from
http://timeslipaccounts.blogspot.com/2009/04/sinking-fee
ling.html/

Thunderbird in the everglades. (2008, August). Retrieved from
http://www.blogcatalog.com/blogs/the-
paranormal-dimension/posts/tag/thunderbird/

Time bus. (2006, August 16). Retrieved from
http://www.ufofreeparanormal.com/stories/viewstory.php
?sid=1427.htm/

Time distortion. (2006, September). Retrieved from
http://paranormal.about.com/library/blstory_september06
_11.htm/

Time glitch in Kansas. (2007, May). Retrieved from
http://paranormal.about.com/library/blstory_may07_02.htm/

Time glitch. (2005, October). Retrieved from
http://paranormal.about.com/library/blstory_october05_1
2.htm/

Time moved too slowly. (n.d.). Retrieved from http://paranormal.about.com/od/timeanddimensiontravel/a/aa072307.htm/

Time phase in the Grand Canyon. (2009, June). Retrieved from http://paranormal.about.com/od/timeanddimensiontravel/a/tales_09_06_21t.htm/

Time slip and watermelon men. (2005, January). Retrieved from http://paranormal.about.com/library/blstory_january05_20.htm/

Time slip on holiday. (2007, August). Retrieved from http://paranormal.about.com/library/blstory_august07_10.htm/

Time slip. (2000, November). Retrieved from http://paranormal.about.com/library/blstory_november00.htm/

Time slips. (2002, May). Retrieved from http://paranormal.about.com/library/blstory_may02_10.htm/

Time—traveling car. (2005, January). Retrieved from http://paranormal.about.com/library/blstory_january05_16.htm/

Torrence, M. (2004, July). *Slip back in time.* Retrieved from http://paranormal.about.com/library/blstory_july04_som.htm/

Toynbee, A.J. (1934-1961). *A study of history* (Vol. 10). Oxford: Oxford University Press.

Traveler. (2007, May). *Missing rummage sale.* Retrieved from http://paranormal.about.com/library/blstory_may07_24.htm/

UnBreakable. (2009, June 6). Retrieved from www.abovetopsecret.com

Vanishing bread. (2009, July). Retrieved from http://paranormal.about.com/od/trueghoststories/a/tales_0 9_07_18t.htm/

von Buttlar, J. (1978). *Time slip—dreams: A parallel reality.* London: Sidgwick & Jackson.

Wagner, S. (2007, July 23). *Out of time and place.* Retrieved from http://paranormal.about.com/od/timeanddimensiontravel/ a/aa072307_2.htm/

Wagner, S. (2007, July 23). *Time moved too quickly.* Retrieved from http://paranormal.about.com/od/timeanddimensiontravel/ a/aa072307.htm/

Wagner, S. (2007, July 23). *When time goes crazy.* Retrieved from http://paranormal.about.com/od/timeanddimensiontravel/ a/aa072307_2.htm/

Wenlong, Y. (2005, June 16). *What does a 200 million year old fossil with a shoe print tell us?* Retrieved from http://www.asianresearch.org/articles/2621.html/

What's in store: Transmuting shop baffles U.K. policeman. (1996, July). Retrieved from http://www.bbc.co.uk/dna/h2g2/alabaster/A684821/

White, E. (2004, August). *He got there before he arrived.* Retrieved from http://paranormal.about.com/library/blstory_august04_15.htm/

Winters, H. (1998, January 25). *Victorian house.* Retrieved from http://www.ufofreeparanormal.com/stories/viewstory.php?sid=334.htm/

Winters, S. (1980). *Shelley: Also known as Shirley.* New York: Ballantine Books.

Wormhole on earth? (2006, September). Retrieved from http://paranormal.about.com/library/blstory_september06_24.htm/

Xylophobia, I. (2008, June 8). *Time slips—Are people really going back in time?* Retrieved from http://www.abovetopsecret.com/forum/thread348068/pg2#pid4755241.htm/

Yknot1. (1995, April). *Time slipage?* Retrieved from yknot1@aol.com/

About the Author

Anita Holmes' vocation is teaching; her avocation, writing! Her writing interests range from history to mystery, with time-slips a priority. She has three degrees (B.S. in Management from Marylhurst, a dual B.S. degree in Education and History from S.U.U., and a Master of Public Administration from Lewis and Clark.)

A widow, she has six children of whom she is inordinately proud, and 13 beloved grandchildren. Holmes is an Oregonian by birth, an Alaskan by profession, and summer of 2011 will be a Missourian by choice.

Holmes looks forward to compiling additional volumes on the remarkable topic of time-slips.

Other Books Published
by
Ozark Mountain Publishing, Inc.

Continue for more books by Ozark Mountain Publishing, Inc.

For more information about any of the above titles, soon to be released titles, or other items in our catalog, write or visit our website:

OZARK
MOUNTAIN
PUBLISHING
PO Box 754
Huntsville, AR 72740
www.ozarkmt.com
1-800-935-0045/479-738-2348
Wholesale Inquiries Welcome